Contents

AS-Level Mathematics modules   4
How to use this book for revision   5
Revision tips   6
Examination technique   7
How this book covers the exam specifications   10

Pure Maths modules
Contents list   14
Revision materials and exercises   15
Answers and hints to solutions   75

Statistics modules
Contents list   78
Revision materials and exercises   79
Answers and hints to solutions   117

Mechanics modules
Contents list   120
Revision materials and exercises   121
Answers and hints to solutions   159

Decision and Discrete Maths modules
Contents list   161
Revision materials and exercises   162
Answers and hints to solutions   189

Index   191

AS-Level Mathematics modules

AS-Level Mathematics gives you more flexibility as a candidate than almost any other AS-Level.

All the major exam boards divide their courses into modules which are classified as Pure Maths, Statistics, Mechanics, or Decision and Discrete Maths. Although different boards use slightly different titles, for simplicity here we call the modules:

P1, P2, P3, … etc.   for Pure Maths
S1, S2, S3, … etc.   for Statistics
M1, M2, M3, … etc. for Mechanics
D1, D2, D3, … etc.   for Decision and Discrete Maths

There is most disagreement among boards about the official title of the 'D' modules. Just count them as D1, D2, etc. for revision purposes.

Secondly, every module is classified as either an AS module, or an A2 module. This is some measure of the relative difficulty of the module, as A2 is more advanced than AS, but it also indicates in some cases the 'dependency' of the modules. In every case, you must complete three modules for an AS qualification, but depending on your choice of examination, some of them may be A2 modules.

Dependency
For example, in all boards, P3 is an A2 module, and you cannot study it without previously studying P1 and P2.

The flexibility of AS-Level Maths means that you can combine different modules to gain different final qualifications. You could aim for:

The titles vary slightly between boards, and not all titles are available from every board. AS Mechanics and AS Decision Maths are not common.

AS Mathematics   AS Mechanics
AS Pure Mathematics   AS Decision Mathematics
AS Statistics   AS Applied Mathematics

The number of different possible combinations from all boards for the full set of possible qualifications is very large!

As well as the P, S, M and D modules, some boards include other modules with other titles in possible combinations. You will need to check your course specification for details.

In this book we have included all the important results and methods for each different version of AS Mathematics. This means that some of the work in this book is from A2 modules, especially in Pure Maths, but the major part of the book covers work from the AS modules.

All AS Mathematics qualifications require that you study two 'pure' modules, and one 'applied' module.

One board includes a 'Methods' module instead of one Pure Maths module. The necessary results are included in the Pure Maths section of this book.

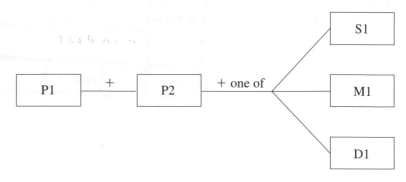

For other AS-Level qualification titles you will need to supplement your revision from other texts.

How to use this book for revision

This book is divided into four sections: Pure Maths, Statistics, Mechanics, and Decision and Discrete Maths. The text gives you the key facts and results, in some cases with the underlying 'proof' or justification for the results.

> Try to learn thoroughly all the key facts and results that are highlighted like this.

To keep the book at a manageable size, we have omitted many proofs which we believe you will be unlikely to be asked to reproduce in an exam.

Each section has a detailed contents page. Because of the different combinations of modules possible in AS-Level, each of the sections contains work that you will not need for your particular specification. Check your course specifications in the tables on pages 10–13 to make sure you know which sections you need to revise.

Worked examples help to show you how the important results and methods are applied to exam-type questions. They have been carefully chosen to include very typical questions, common methods that you must know, and common applications.

Throughout the book, notes in tint boxes give you extra hints and help. Each section of work provides typical questions for you to try. Make sure you work through them.

Complete answers and hints for solutions are given at the end of each section. Don't short-cut this work – questions reinforce what you know better than just reading text. As you become more confident, find past papers and practise a 'real' exam too.

Remember your GCSE work

All AS-Level specifications make assumptions about work you have done at GCSE, before you start the AS course.
We have tried not to assume too much, and we have included some material which is not, strictly, AS-Level work. But we have assumed, for instance, that you can factorise quadratic expressions, and can draw Cartesian graphs. We have also assumed that you own and can use a scientific calculator.

> Leaving out some proofs doesn't mean that we don't think they are important! We do. But this is *not* the moment to try to understand and learn them for the first time.

> Look at these carefully – they should help to fill in any gaps in your understanding. Sometimes they explain further some underlying point, but we have been careful not to introduce any completely new ideas. Other notes remind you of key exam hints, or revision points.

Revision tips

Keep these points in mind while you revise.

- Know what the exam consists of. If you are not at school or college, you can order the **Specifications** from the exam board. For this you need to know your exam board address and the syllabus specification number.

- It is important to have a selection of past papers, but make sure you have the most recent papers. Ask your school or college, or the exam board, if there have been any changes to the syllabus specification for your examination year. If so, make sure you have a copy of the new **specimen paper**.

- Study the past papers and familiarise yourself with the layout of the paper.

- Draw up a revision plan for the weeks before your exam, and stick to it. Don't convince yourself that you can 'catch up' everything you have missed in the last few days before the exam.

- Revise in short manageable chunks. Do not attempt to do all of the subject in one go, but take each topic in turn, working to your revision plan.

- When you revise, use whichever method or methods you feel most happy with. These could include making notes, practising diagrams, explaining out loud to a friend, reading your textbook, and many other things, as well as working through past papers. For most people, a combination of methods is best. Reward yourself with a treat! (But be honest – make sure you deserve it!)

- Take regular breaks. It is difficult to concentrate for more than about 40 minutes at a time. Have a 15-minute break between sessions. Take a walk, or do something else, during your breaks.

- Test yourself. Ask a teacher, parent or friend to assess you. If you ask a friend, then two of you are revising and helping each other's work. This is very often the best method.

- Do not work too late, get plenty of sleep, and try not to get too bored with revision.

Examination technique

There are a few general principles that apply to AS-Level Maths questions, whatever the type of question, or the style of your paper.

Read the rubric of the paper, usually on the front page. **Make sure that you answer the correct number of questions** and always check, because the exam style may have changed and you may have missed the information. If the paper has sections, **choose the correct number from within each section**.

Read the rubric.

Check the **rubric** for standard rules. In mechanics, some exam boards insist on a particular approximation for g, the acceleration due to gravity. You will often find things like: 'Use $g = 9.8$, unless otherwise stated in the question.' Don't then use something different.

Get to know what is in your **formula book**, if there is one. Don't waste time in an examination trying to work out a half-remembered formula, when it is clearly stated in your formula book.

Know your formula book.

Always show your working in a question, unless it explicitly states that marks will be given only for the correct answer. Usually you will receive credit for working, even if it leads to the wrong answer. If you are short of time, you can sometimes interpret where credit will be given only for the correct answer. The key is to look for phrases such as 'Write down the number of ...'.

Show your working.

Don't arrive at the examination unprepared. Make sure:

- **your calculator has a set of fresh batteries**
- **you know how your calculator works**
- **you have your formula booklet (usually one will be supplied by the Examination Centre – check beforehand)**
- **you have pens, a pencil, pencil sharpener, ruler and eraser**
- **you have the basic drawing instruments if there is any possibility that you may need to do a scale drawing.**

Take some simple, defensive, precautions.

Success in examinations

If you have put in a lot of effort for a year or two, there are some steps you can take to ensure that your examination result reflects it.

Be prepared

If the thought of an examination fills you with dread, then you are probably not fully prepared. Of course, everyone is nervous about an important examination, but there is no reason why you should fall apart and fail to do yourself justice if you know what to expect before you go into the exam room. The best preparation is working through past papers and looking at your syllabus specifications to avoid any nasty surprises. A positive attitude will help you too – try to look on the examination as an opportunity to show the examiner all the things you have learned in the past year or two.

Make sure you have everything you need ready the night before.

Be prepared.

Read carefully

Before you start the exam, look over all the pages just to check that the paper is complete and in the format you expect, so that you do not miss any parts. Read the rubric, and follow the instructions to the letter.

Plan your time

Work out how much time you should spend on each question, based on how many marks it has. Don't spend valuable time trying to think of the answer to the last remaining part of a question when you don't really understand it – go on and start the next question instead. In AS-Level mathematics, if you find yourself filling a couple of sides of A4 paper with algebra, you have almost certainly followed the wrong method. Leave it and go back later, if you have time left at the end.

Present your work clearly

Remember that the examiner will have a large pile of papers to mark in a short space of time. Struggling to read your algebra or to follow your method will not help, so write as accurately as you possibly can in the time available. The examiner is trying to award you marks – make it easy to find them.

Diagrams

When you draw diagrams, keep them clear and simple. Label them carefully, using consistent conventions. For instance, distinguish between forces, velocities and accelerations in mechanics questions by using different types of arrow.

Crossing out

If you cross out some working or part of your working when you have finished, then do so with one clear 'crossing-out' line. You may be surprised to know that candidates in maths exams are frequently credited with marks for things they have crossed out, so do not obliterate them with a scribble, and do hand them in. Some exam boards are very generous to candidates, but if you have for some reason attempted a question twice, do be clear about which version you want marked, or usually it will be 'first attempt counts'.

Stay calm

If you find a question you have no idea about, do not panic! Breathe slowly, and have another look. It's likely that if you stay calm and think clearly it will start to make more sense, or at least you may be able to answer part of it. If not, then don't agonise about it but concentrate first on the questions you can answer.

Question types

Your examination may contain several of the same type of question, or it may contain different styles of question. Calculate the time allocation to individual questions on the basis of the marks given. Don't be too rigid in sticking to the time, but don't, for example, spend two-thirds of the time trying to earn one-quarter of the marks.

Short-answer questions

These questions usually test only one principle or idea. Usually you will need to produce a diagram or a few lines of mathematics. The number of marks is always given, and is a guide to the level of detail required. Don't create pages of working, or use lots of valuable time on a question that is worth only a few marks.

> Try to spot where the question just asks for the answer to be written down. It must be easy then.

Structured questions

These are longer questions made up of connected parts. They test your understanding of the development of ideas in mathematics, often linking together associated kinds of problem. The concept of creating a **mathematical model**, and interpreting a result from it, is often tested in extended structured questions. The number of marks is given on the paper for each part of the question.

Answering the questions

Make sure you read the question carefully, and do not answer a different question that you expected to be asked or which you wanted the examiner to ask. The number of marks gives a good indication of how much detail the examiner is looking for.

Don't despair if the first part of an extended question puzzles you. Examiners are good at structuring questions so that candidates are not penalised by losing all the marks just because they cannot do one part. The question will usually contain information to allow you to continue with later parts, even if you have an early part wrong, or which you cannot do.

> Get some help from the Examiner.

In many questions, a diagram is likely to be helpful. Don't draw anything but a freehand sketch unless the question asks you either to draw a **scale diagram** (unusual, but not unknown, in AS-Level) or a **graph** (which is common in Statistics modules). If you do draw a precise graph, remember that you will be assessed on its quality, so draw it in (sharp) pencil, and label your axes accurately and sensibly. Use a ruler where appropriate, and be bold (not fuzzy) when joining plotted points to draw freehand curves.

> Make your diagrams earn you marks.

Don't be afraid to write a sentence or two in English to explain your reasoning to an examiner – you will receive credit if it helps explain your method. In some questions, particularly in Statistics and modelling, you may be asked to write a brief interpretation of your results. In such an instance, be explicit and keep it short. The number of marks available is a hint of what to write: if there are, say, 2 marks for a brief explanation, the examiner is likely to be looking for 2 important points – he or she may have a list which gives the marks for 'any 2 from these 5' or something similar. If both points can be explained in one brief sentence, you will still get all the marks. One of the most common errors is to say something twice, in two slightly different ways, so that the examiner cannot award the marks for either.

> Write sense.

> Use the marks given as a guide.

How this book covers the exam specifications

These tables show where the topics given in the different exam boards' specifications are covered in this *AS-Level Maths Revision Guide*.

AS-Level Maths – Pure Maths modules

Chapter	Topic	OCR (MEI)	OCR A	AQA A	AQA B	Edexcel
1	**Functions and mappings**					
	Composite functions	P2	P2	P1	P1	P2
	Inverse functions	P2	P2	P1	P1	P2
	Odd, even and periodic graphs	P2		P1	P2	P2
	Sketching graphs	P1/P2	P2	P1	P1/P2	P2
	Graphs and transformations	P2	P1	P1	P1	P2
	Asymptotes and behaviour for large values of x	P1/P2	P1	P2	P3	P2
2	**Coordinate geometry – straight lines**					
	The equation of a straight line	P1	P1	Me	P1	P1
	Midpoint, and distance between two points	P1	P1	Me		P1
3	**Algebra and series**					
	Quadratic equations	P1	P1	Me	P1	P1
	Powers and indices	P2	P1	Me	P1	P1
	Polynomials	P1	P2	Me	P2	P1
	The factor theorem	P1	P2	Me	P2	P1
	Series: Arithmetic progressions	P2	P2	P1	P1	P1
	Geometric progressions	P2	P2	P1	P2	P1
4	**Trigonometry**					
	Trigonometric functions	P1	P1	P1	P1	P1
	Inverse trigonometric functions	P1	P3	P1	P1	P1
	The sine rule and the cosine rule	P1	P1			P1
	Radians – circular measure	P1	P2	P1	P2	P1
	Arc length and the area of a sector	P1	P2	P1	P2	P1
	Trigonometric formulae – compound and double angles	P3	P3	P2	P4	P2
	Using $a \cos x + b \sin x$	P3	P3	P2	P4	P2
5	**Calculus – understanding change**					
	Basic differentiation	P1	P1	Me/P1	P1	P1
	Tangents and normals to a graph	P1	P1	Me/P2	P4	P1
	Stationary values – maxima and minima	P1	P1	Me/P1	P1	P1
	Second derivatives	P2	P1	P1	P2	P1
	Points of inflexion	P1		P1	P1	P1
	Basic integration	P1	P1	Me/P1	P1	P1
	Area between a graph and the x-axis	P1	P1	Me/P1	P1	P1
	Area between two curves	P1	P1			
	Volume of a solid of revolution	P1	P2	P2	P4	P2
6	**Using differentiation**					
	The chain rule – a function of a function	P2	P2	P2	P4	P3
	Linked rates of change	P2	P2		P2	
	Integration using the chain rule	P2	P2	P2	P4	P3
	The product rule	P2	P3	P2	P4	P3
	The quotient rule	P2	P3	P2	P4	P3
7	**Logs and exponentials**					
	Indices and the laws of logarithms	P2	P2	P1	P2	P2
	Exponential functions	P2	P2	P1	P2	P2
	The function e^x	P2	P2	P1	P2	P2
	Negative exponential functions	P2	P2	P1	P2	P2
	Exponential and logarithm functions	P2	P2	P1	P2	P2
	Exponentials, logs and calculus	P2	P2	P1	P2	P2
8	**Numerical methods**					
	Solving equations by numerical methods	P2	P2	P1	P1	P2
	The decimal search method	P2	P2	P1	P1	P2
	Interval bisection method	P2		P1	P4	
	Fixed point methods – formula iteration	P2	P2		P4	P2
	Convergence of formula iteration	P2	P2		P4	P2
	The Newton–Raphson method	P2			P4	
	Numerical integration – the trapezium rule	P1	P2			P2

AS-Level Maths – Statistics modules

Chapter	Topic	OCR (MEI)	OCR A	AQA A	AQA B	Edexcel
1	**Discrete data**					
	Measures of centre and spread	S1	S1	Me	S1	S1
	Stem and leaf diagrams	S1	S1	Me	S1	S1
	Box plots	S1	S1	Me	S1	S1
2	**Continuous data**					
	Measures of centre	S1	S1	Me	S1	S1
	Variance and standard deviation	S1	S1	Me	S1	S1
	Histograms	S1	S1	Me	S1	S1
	Cumulative frequency graphs	S1	S1	Me	S1	
	Linear coding	S1	S1	Me	S1	
3	**Orderings**					
	Factorials; combinations and permutations	S1	S1			
	Arranging similar objects		S1			
	Sampling without replacement	S1	S1			
	Sampling with replacement	S1	S1			
4	**Probability**					
	Probability scale and definitions	S1	S1	Me	S1	S1
	Union, intersection, complement	S1	S1	Me	S1	S1
	Mutually exclusive events	S1	S1	Me	S1	S1
	Independence	S1	S1	Me	S1	S1
	Conditional probability: formula	S1	S1	Me	S1	S1
	Independence with conditional probability	S1	S1	Me	S1	S1
5	**Discrete random variables**					
	Definition of discrete random variable		S1	Me/S1		S1
	Expectation and variance		S1	Me/S1		S1
	Cumulative distribution function					S1
6	**Discrete probability distribution**					
	Binomial: rule, expectation and variance	S1	S1	S1	S1	
	Using cumulative tables	S1	S1	S1	S1	
	Expected frequency	S1		S1		
	Hypothesis testing	S1				
	Poisson: rule, expectation and variance				S1	
	Using cumulative tables				S1	
	Sums of independent Poisson distributions				S1	
	Geometric: rule, expectation and variance		S1			
	Discrete uniform: rule, expectation and variance					S1
7	**Expectation algebra**					
	Expectation, variance					S1
	Expectations of functions					S1
	Rules for linear functions					S1
8	**Continuous random variables**					
	Probability density function			S1		
	The normal distribution: standard normal curve			S1	S1	S1
	Using tables			S1	S1	S1
	Other normal distributions			S1	S1	S1
	Finding μ and σ			S1	S1	S1
9	**Sampling**					
	Sampling frames and sampling methods	S1		S1	S1	
10	**Bivariate data**					
	Scatter diagrams		S1		S1	S1
	Pearson's product moment		S1		S1	S1
	Linear regression		S1		S1	S1

AS-Level Maths – Mechanics modules

Chapter	Topic	OCR (MEI)	OCR A	AQA A	AQA B	Edexcel
1	**Distance, velocity and acceleration**					
	Scalars and vectors	M1	M1	M1	M1	M1
	Addition and subtraction of vectors	M1	M1	M1	M1	M1
	Resolving vectors – components	M1	M1	M1	M1	M1
	Constant acceleration formulae	M1	M1	M1	M1	M1
	Constant acceleration and graphs	M1	M1	M1	M1	M1
	Calculus and kinematics	M1	M1	M1	M1	M2
2	**Forces and acceleration**					
	Newton's laws of motion	M1	M1	M1	M1	M1
	Mass and weight	M1	M1	M1	M1	M1
	The equation of motion – Newton's second law	M1	M1	M1	M1	M1
	Connected particles – Newton's third law	M1	M1	M1	M1	M1
3	**Equilibrium**					
	Triangle of forces – equilibrium under three forces	M1	M1	M1	M1	M1
	Particle in equilibrium on an inclined plane	M1	M1	M1	M1	M1
4	**Friction**					
	$Fr \leq \mu R$		M1	M1	M1	M1
	Force applied at an angle		M1	M1	M1	M1
	Rough inclined planes		M1	M1	M1	M1
5	**Momentum and impulse**					
	Momentum		M1	M1	M1	M1
	Impulse		M1	M1	M1	M1
	Collisions		M1	M1	M1	M1
6	**Projectiles – motion in a vertical plane**					
	Horizontal projection	M1	M1	M1	M1	M2
	Projection at an angle	M1	M1	M1	M1	M2
	Time of flight, greatest height	M1	M1	M1	M1	M2
	Range of flight	M1	M1	M1	M1	M2
	Equation of trajectory	M1	M1	M1	M1	M2

AS-Level Maths – Statistics modules

Chapter	Topic	OCR (MEI)	OCR A	AQA A	AQA B	Edexcel
1	**Algorithms**					
	Flow charts and written rules	D1	D1	D1		D1
	Bubble sort	D1	D1	D1		D1
	Shuttle sort	D1	D1	D1		D1
	Quick sort	D1		D1		D1
	Binary search	D1				D1
	Bin packing	D1	D1			D1
2	**Graphs and networks**					
	Definitions	D1	D1	D1	D1	D1
	Planar and non-planar graphs	D1	D1		D1	D1
	Bi-partite graphs and matchings			D1		D1
	Flows in networks					D1
	Prim's algorithm	D1	D1	D1	D1	D1
	Kruskal's algorithm	D1	D1	D1	D1	D1
3	**Shortest path problems**					
	Dijkstra's algorithm	D1	D1	D1	D1	D1
	Chinese postman problem		D1	D1		D1
4	**Critical path analysis**					
	Precedence diagrams	D1			D1	D1
	Cascade charts	D1			D1	D1
	Scheduling	D1			D1	D1
5	**Linear programming**					
	Formulation and graph drawing	D1	D1	D1	D1	D1
	Tour of vertices and integer solutions	D1	D1	D1	D1	D1
	The simplex method		D1			D1
6	**Boolean algebra**					
	Propositional logic					D1
	Switching circuits and combinatorial circuits					D1

Pure Maths Topics:

1 Functions and mappings

Definitions	15
Composite functions	15
Inverse functions	16
Odd, even and periodic functions	16
Graphs of functions	17
Using the standard graphs	18
Graphs and transformations	19
Asymptotes and behaviour for large values of x	20
Exercise 1: Functions and mappings	21

2 Coordinate geometry

Equation of a straight line	22
Midpoint of a straight line	23
Distance between two points	23
Exercise 2: Coordinate geometry	24

3 Algebra and series

Quadratic equations	25
Powers and indices	26
Polynomials	27
The factor theorem	27
Series	28
Arithmetic progressions	28
Geometric progressions	29
Exercise 3: Algebra and series	30

4 Trigonometry

Trigonometric functions	31
Inverse trigonometric functions	32
The sine rule and the cosine rule	33
Radians	35
Trigonometric formulae	36
Exercise 4: Trigonometry	37

5 Calculus – understanding change

Gradient of a curve	38
Differentiating algebraic expressions	39
Finding the tangent to a graph	40
Finding a normal to a curve	41
Maxima and minima	42
Points of inflexion	44
Integration	46
Integrating algebraic expressions	46
Indefinite integration	47

Area between a graph and the x-axis	48
Interpreting a negative result for an integral	49
The area enclosed between two curves	50
Volume of a solid of revolution	51
Exercise 5: Calculus – understanding change	52

6 Differentiation and integration

The chain rule	53
Using the chain rule	54
Calculating rates of change for linked variables	54
Integration by inspection using the chain rule	55
The product rule	56
Using the product rule	57
The quotient rule	58
Using the quotient rule	59
Exercise 6: Differentiation and integration	61

7 Logs and exponentials

Definition of a log	62
Indices and the laws of logs	62
Solving problems with logs	62
Exponential functions	63
The exponential function and e	63
Negative exponentials	64
Exponentials and logs	64
Exponentials, logs and calculus	65
Differentiating and integrating exponentials	65
Exercise 7: Logs and exponentials	67

8 Numerical methods

Solving equations by numerical methods	68
Decimal search method	68
Interval bisection method	69
Fixed-point methods	70
Solving equations by formula iteration	70
Why does formula iteration work?	71
The Newton–Raphson method	72
Numerical integration – the trapezium rule	73
Exercise 8: Numerical methods	74

Answers and hints to solutions	75

1 Functions and mappings

Definitions

A **mapping** is a rule connecting the elements of two sets or, sometimes, the elements of the same set.

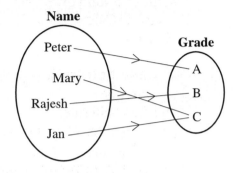

In most cases at AS-Level, the elements of the sets are numbers, and the mapping is an algebraic formula.

A mapping can be one-to-one, many-to-one, one-to-many, or many-to-many.

The set being mapped *from* is the **domain**, and the set being mapped *to* is the **codomain**. The elements of the codomain which are actually mapped to are called the **range**. The range could be a subset of the codomain, or it could be the whole codomain.

The important word here is **range** – the outputs you actually get from the mapping.

If every element of the domain maps to exactly one element of the codomain, then the mapping is called a **function**. So a function must be one-to-one, or many-to-one.

If a mapping is a function, every value you put in gives you one value out.

Composite functions

When one function is applied to the result of another, a **composite function** is formed.

EXAMPLE

If $f(x) = 2x$ and $g(x) = 3x + 1$
both with domain and codomain real numbers, find a composite function $h(x) = gf(x)$.

$$f(x) = 2x \Rightarrow gf(x) = g(2x)$$
$$= 3(2x) + 1$$
$$= 6x + 1$$

So the composite function $h(x)$ is given by
$$h(x) = 6x + 1$$

The domain of $h(x)$ is also the real numbers. Any real number can be substituted in $f(x)$, and then the outcome can be substituted into $g(x)$.

REVISION NOTE

The real numbers include all integers, rational and irrational numbers. They are often denoted by $\mathbb{R}$.

REMEMBER

For composite functions, make sure that the 'output' of the first function, its range, is part of the domain of the second.

The composite function $h(x) = gf(x)$ means 'first apply function f to x and then apply function g to the outcome'.

So it represents the effect of two functions applied consecutively.

Take care not to confuse the composite function $gf(x)$ with the product of the two functions $g(x).f(x)$.

REVISION NOTE

Notice the order. The composite function gf means 'do f first and then do g'.

To reinforce that it's something special, some people use $g \circ f$ to stand for the composite function.

Inverse functions

> If a function $f(x)$ is one-to-one, then there is an inverse function $f^{-1}(x)$.

So if $y = f(x)$ is a mapping that gives a value of y for a given x, then $x = f^{-1}(y)$ is an inverse mapping which 'undoes' the effect of the first mapping.

> The graph of an inverse function is a reflection in $y = x$ of the graph of the original function.

For $y = \dfrac{1}{x+1}$ the function is undefined for $x = -1$. In the same way, there is no value of x which gives $y = -1$ for the inverse.

The diagram shows $y = \dfrac{1}{x+1}$, its inverse function in colour, and the line of symmetry $y = x$.

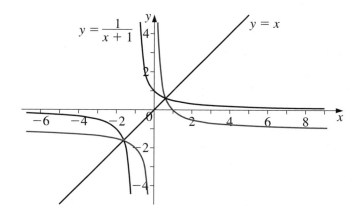

$y = \dfrac{1}{x+1}$ $y = x$

Odd, even and periodic functions

For any value of x:

> If a function is even, then $f(x) = f(-x)$ and the y-axis is a line of symmetry on the graph.
>
> If a function is odd, then $f(x) = -f(-x)$ and the graph will map onto itself with a half-turn about the origin.

$y = x^4 - 2x^2 - 1$
even

$y = x^3 - x$
odd

> If a function is periodic, then $f(x + c) = f(x)$ for every x and a constant c, so the graph shape repeats after a step along the x-axis of length c.

Graphs of functions

A graph provides a simple way of understanding the behaviour of a function. You need to know and understand the graphs of some basic functions.

The straight line

> The graph of a straight line is given by
>
> $$y = mx + c$$
>
> where m is the gradient of the line, and c is the intercept on the y-axis.

There is more about straight lines on page 22.

Standard curves

This is the graph of $y = x^2$.

The graph (and function) is:

(i) 'even' because $f(x) = f(-x)$
(ii) a parabola
(iii) sometimes called a 'quadratic' graph.

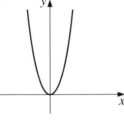

This is $y = x^3$.

The graph is 'odd' because $f(x) = -f(-x)$ and is called a 'cubic' because the highest power is x^3.

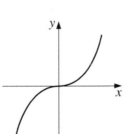

REMEMBER

- Positive gradients go 'up' to the right
- Negative gradients go 'down'
- Two lines are perpendicular if
$$m_1 = -\frac{1}{m_2}$$

REVISION NOTE

You need to be able to find where straight lines cross, and calculate some distances – see page 23.

A **quadratic** graph is always a parabola, and a **cubic** graph has a characteristic 'S' shape. The sign of the highest power determines the 'direction' of the graph.

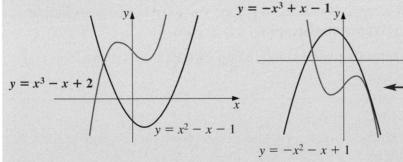

$y = x^3 - x + 2$
$y = x^2 - x - 1$
$y = -x^3 + x - 1$
$y = -x^2 - x + 1$

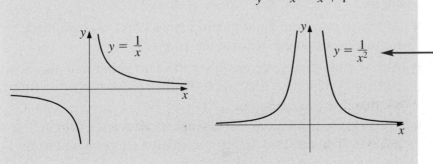

$y = \frac{1}{x}$
$y = \frac{1}{x^2}$

REMEMBER

Higher powers have more bends. For example, if the highest power is degree 4, then the shape is a deformed 'W'.

In the graphs on the left, as $x \to +\infty$, $y \to +\infty$.
In the graphs on the right, as $x \to +\infty$, $y \to -\infty$.

For both graphs, there is no value of y when the value of x is 0.

For both graphs, the axes are **asymptotes** – lines that the graphs approach more and more closely, but never touch.

Using the standard graphs

You can sketch more complicated graphs by combining standard graphs.

EXAMPLE

Draw the graphs of $y = x$ and $y = \dfrac{1}{x^2}$ on the same sketch and combine them to draw $y = x + \dfrac{1}{x^2}$.

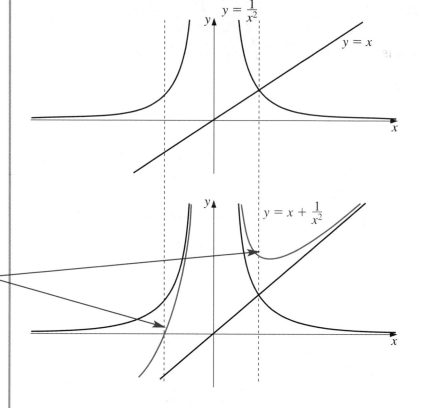

You need to be able to sketch these graphs quickly, without plotting points exactly – learn the shapes.

There are other standard graphs you need to know, including the **Exponential graphs** – see Section 7.

Other important methods using **Calculus** are covered in Section 5.

METHOD NOTE

First draw the two separate graphs.

Now, graphically 'add them up' to get the graph you want.

Notice that the original graphs are 'asymptotes' for the new graph.

The two original graphs 'cancel each other out' and 'add-up' at these points.

REMEMBER

Graphical calculators are not allowed in some exams.

Use all the evidence you can collect from the separate graphs. For example:

- At any point where the standard graphs cross, a graph made from their difference will cut the x-axis.

- Any vertical asymptote on either standard graph will also be a vertical asymptote if that graph is added to or subtracted from another.

- The behaviour for large x of a graph made from a sum or difference of standard graphs can easily be interpreted from the behaviour of the standard graphs.

Graphs and transformations

Sometimes you can sketch the shape of a graph by understanding how it is related to a simpler graph.

Each graph below is a transformation of the graph of $y = f(x)$ where $f(x) = x^3 - x + 1$

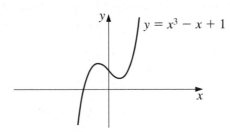

$y = x^3 - x + 1$

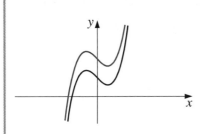

$y = f(x) + k$

The graph is a translation of k units upwards

i.e. a translation by $\begin{pmatrix} 0 \\ k \end{pmatrix}$.

Here, $k = 1$.

> In the first two transformations the shape of the graph is unchanged.

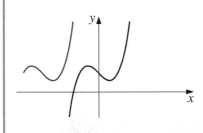

$y = f(x + k)$

The graph is a translation of k units to the left

i.e. a translation by $\begin{pmatrix} -k \\ 0 \end{pmatrix}$.

Here, $k = 3$.

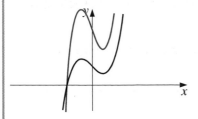

$y = \alpha f(x)$

The graph is a stretch parallel to the y-axis, scale factor α.

Here, $\alpha = 3$.

> In this case, each y-coordinate is multiplied by 3. Of course, the intercept on the x-axis is unchanged.

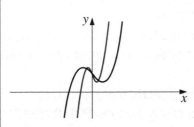

$y = f(\alpha x)$

The graph is a stretch parallel to the x-axis of scale factor $\dfrac{1}{\alpha}$.

Here, $\alpha = 2$.

> Here the graph is 'squashed' in the x-direction. In this case, the intercept on the y-axis is unchanged.

For example, for $x = -100$, you get $y = \dfrac{-199}{-102} = 1.9509\ldots$ or just less than $+2$.

For $x = -1000$, it's even closer to 2.

Sketching graphs

1. Check for standard functions.
2. Check for intercepts on the axes.
3. Check for symmetry – even, odd, periodic.
4. Look for vertical asymptotes.
5. Look for behaviour of large x, positive and negative.

Asymptotes and behaviour for large values of x

For some functions it is easy to spot values for which the function is undefined. These will often give values which appear on the graph as a **vertical asymptote**.

For example, for the graph $y = \dfrac{2x+1}{x-2}$, the value $x = 2$ makes the denominator zero and hence makes the function undefined.

Another useful method to use when sketching graphs of functions is to check the **behaviour for very large values of x** (positive and negative). This will sometimes help you to identify a **horizontal asymptote**.

For the graph of $y = \dfrac{2x+1}{x-2}$, by inspection

$$\text{as } x \to +\infty, y \to 2 \ \text{(from above)}$$
$$\text{and as } x \to -\infty, y \to 2 \ \text{(from below)}$$

so the line $y = 2$ will be a horizontal asymptote.

Always try to find where the graph will cross the axes.

When $x = 0$, $y = \dfrac{0+1}{0-2} = -\dfrac{1}{2}$, so the graph crosses the y-axis at $(0, -\frac{1}{2})$

When $y = 0$,

$$0 = \frac{2x+1}{x-2} \ \Rightarrow \ 2x+1 = 0 \ \Rightarrow \ x = -\tfrac{1}{2}$$

so the graph crosses the x-axis at $(-\frac{1}{2}, 0)$

Plot the two intercepts with the axes and the two asymptotes. It is easy to fill in a sketch of the graph of the function.

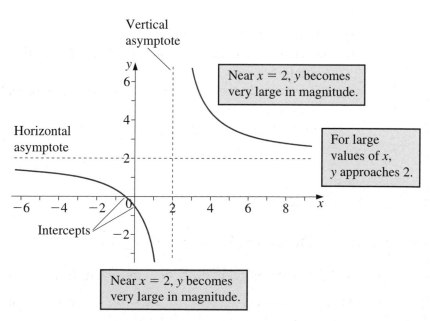

Vertical asymptote

Horizontal asymptote

Near $x = 2$, y becomes very large in magnitude.

For large values of x, y approaches 2.

Intercepts

Near $x = 2$, y becomes very large in magnitude.

Functions and mappings

1 (a) If $f(x) = 3x + 1$ and $g(x) = x^2$, where the domain of both mappings is $\mathbb{R}$, the real numbers, find

(i) $f \circ f$

(ii) $f \circ g$

(iii) $g \circ f$

For each, write down the range of the composite function.

(b) Which of the composite functions in (a) has an inverse?
Find $h(x)$ the inverse function, in this case.

2 (a) Sketch the graphs of $y = x + 2$, and $y = \dfrac{1}{x}$.

(b) By using simple transformations of graphs, sketch the graph of

$$y = \frac{3}{x - 2},$$

identifying the vertical asymptote.

(c) Show that

$$(x + 2) + \frac{3}{x - 2} = \frac{x^2 - 1}{x - 2}$$

and by using the sketches you made in (a) and (b), sketch the graph of $y = \dfrac{x^2 - 1}{x - 2}$.

3 (a) For the graph

$$y = \frac{4x - 3}{2x + 5}$$

find:

(i) where the graph crosses the axes,

(ii) the value of x where y is undefined, and hence the equation of the vertical asymptote,

(iii) by inspection, the value of y that the curve approaches for large positive and large negative values of x.
Hence write down the equation of the horizontal asymptote.

(b) Using the information from (a), sketch the graph of

$$y = \frac{4x - 3}{2x + 5}.$$

4 (a) Sketch the graphs of $y = \dfrac{1}{x}$ and $y = \sqrt{x}$.

(b) Find the crossing point of the two graphs in (a).

(c) Use the graphs in (a) and the coordinates of the crossing point to sketch

$$y = \frac{1}{x} + \sqrt{x}$$

2 Coordinate geometry

Equation of a straight line

Any straight line graph can be represented by an equation of the form

$$y = mx + c, \text{ where } m \text{ and } c \text{ are constants}$$

except for lines parallel to the y-axis.

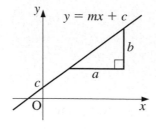

m **is the gradient** of the straight line, and c **is the intercept** (the crossing point) on the y-axis.

On the graph shown, the gradient is

$$m = \frac{b}{a}$$

EXAMPLE

Find the equation of a straight line through $(1, 3)$ parallel to $y = 2x - 1$.

Since the new line is parallel to $y = 2x - 1$, both lines have gradient 2. So the new line has equation

$$y = 2x + c$$

Since this new line passes through $(1, 3)$, these coordinates can be substituted into the equation to find c.

Substituting, $3 = 2 \times 1 + c$

so $\qquad c = 1$

and the new equation is $y = 2x + 1$

EXAMPLE

Find the equation of the straight line through the points $(-2, 5)$ and $(4, -4)$.

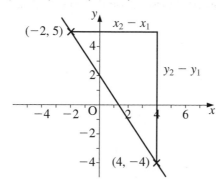

Firstly, find the gradient of the required line. From the diagram, the gradient can be found by subtracting the y-coordinates and subtracting the x-coordinates to find the distances marked.

$$m = \frac{y_2 - y_1}{x_2 - x_1} = \frac{(-4) - 5}{4 - (-2)} = -\frac{3}{2}$$

So the equation is $y = -\frac{3}{2}x + c$, and this passes through the point $(-2, 5)$, so $5 = -\frac{3}{2} \times (-2) + c \Rightarrow c = 2$.

So the equation is $y = -\frac{3}{2}x + 2$.

Midpoint of a straight line

> If a straight line joins the two points (x_1, y_1) and (x_2, y_2) then the midpoint of the line is the point
> $$\left(\frac{x_1 + x_2}{2}, \frac{y_1 + y_2}{2} \right)$$

Distance between two points

> The distance between the two points (x_1, y_1) and (x_2, y_2) is given by
> $$d = \sqrt{(x_2 - x_1)^2 + (y_2 - y_1)^2}$$

REVISION NOTE

The distance, of course, is given by a *positive* square root.

In AS-Level examinations, these results are likely to turn up in questions that test your basic geometry.

EXAMPLE

The line $5x + 12y = -17$ passes through P(11, −6) and the line $12x - 5y = -7$ passes through Q(4, 11). Find point R, where the lines cross, and show that triangle PQR is right-angled and isosceles.

The intersection R is given by solving simultaneously

$$\left. \begin{array}{l} 5x + 12y = -17 \\ 12x - 5y = -\ 7 \end{array} \right\} \Rightarrow \left. \begin{array}{l} 60x + 144y - -204 \\ 60x -\ \ 25y = -\ 35 \end{array} \right\}$$

Multiplying the first equation by 12, and the second by 5…

$$\Rightarrow 169y = -169 \Rightarrow \left. \begin{array}{l} y = -1 \\ x = -1 \end{array} \right\}$$

… and then subtracting to eliminate x.

So R is the point $(-1, -1)$.

The simplest way to investigate the triangle PQR is to find the gradients of its sides. When you find two perpendicular sides, they enclose the right angle.

$$\text{gradient PQ} = \frac{11 - (-6)}{4 - 11} = -\frac{17}{7}$$

$$\text{gradient QR} = \frac{(-1) - 11}{(-1) - 4} = \frac{12}{5}$$

$$\text{gradient RP} = \frac{(-6) - (-1)}{11 - (-1)} = -\frac{5}{12}$$

So QR is perpendicular to RP. If an isosceles triangle contains a right angle, the two sides enclosing the right angle must be equal, so we must check QR and RP.

$$QR = \sqrt{((-1) - 4)^2 + ((-1) - 11)^2} = \sqrt{(-5)^2 + (-12)^2} = 13$$

$$RP = \sqrt{(11 - (-1))^2 + ((-6) - (-1))^2} = \sqrt{12^2 + (-5)^2} = 13$$

So QR = RP, and the angle at R is a right angle.

The diagram looks like this:

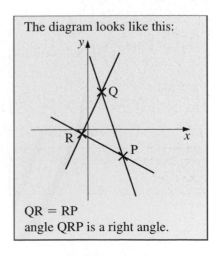

QR = RP
angle QRP is a right angle.

Coordinate geometry

1 Find the equations of the straight lines

(a) line l_1: through A(−1, 2) with gradient $-\frac{1}{2}$

(b) line l_2: through A(−1, 2) and B(4, −2)

Show that C(5, −1) lies on the line l_1, and find the equation of a straight line l_3 through C, which is perpendicular to l_2.

Hence, by finding where line l_3 meets line l_2, find the shortest distance from C to AB.

2 The points A(−1, 1) and B(7, −1) form the base of a right-angled triangle. The hypotenuse of the triangle passes through A, with gradient $\frac{1}{2}$.

Find the coordinates of C, the third vertex of the triangle, and hence the length of the hypotenuse.

3 The position of the centre of mass of a triangular lamina is known as the centroid of the triangle. The centroid lies at the intersection of the 'medians'. Each median of a triangle is a straight line joining a vertex to the midpoint of the opposite side of the triangle.

A triangle ABC has A(6, 30), B(0, 6) and C(24, 0).
Find the coordinates of the centroid of the triangle ABC as follows:

(a) Find the coordinates of M, the midpoint of AC, and of L, the midpoint of BC.

(b) Find the equation of the median AL.

(c) Find the equation of the median BM.

(d) Find the centroid G, the point of intersection of the medians AL and BM.

(e) Check that G lies on the third median.

4 A(−3, 2) and B(4, −1) form the base of a triangle.

(a) Find the length of AB.

(b) The third vertex of the triangle is at C. If angle BAC is a right angle, find the equation of AC.

(c) If the triangle is isosceles, find the length BC.

5 ABCD is a parallelogram, where A is the point (1, 1) and D is the point (2, 5).

(a) The sides AB and CD each have gradient $\frac{1}{2}$. Find the equations of AB and CD.

(b) Side BC has equation

$$y = 4x - 17$$

Find the coordinates of B and C.

(c) Find the lengths of the diagonals of the parallelogram.

3 Algebra and series

Quadratic equations

A **quadratic equation** can always be written as

$$ax^2 + bx + c = 0$$

and the quadratic expression on the left-hand side can be drawn on a graph of $y = ax^2 + bx + c$.

a positive

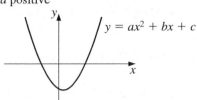

$y = ax^2 + bx + c$

a negative

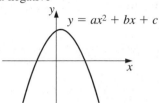

$y = ax^2 + bx + c$

The sign of '*a*' in the expression determines whether the graph 'points down' or 'points up'.

If the graph cuts the *x*-axis, then those values of *x* are the solutions, called the **roots of the equation**.

You should be able to solve a quadratic equation by three different methods:

- Factorisation
- Completing the square
- Using the quadratic formula
$$x = \frac{-b \pm \sqrt{b^2 - 4ac}}{2a}$$

REVISION NOTE

Practise all 3 methods, and make sure you remember the formula.

EXAMPLE

Solve $2x^2 = 13x - 15$

The equation can be written in the usual format as

$$2x^2 - 13x + 15 = 0$$
$$\Rightarrow (2x - 3)(x - 5) = 0$$
$$\Rightarrow \text{ either } x = \tfrac{3}{2} \text{ or } x = 5$$

Factorising is a 'trial and error' process, but notice that the middle term comes from
$$2 \times (-5) + (-3) \times 1 = -13$$

EXAMPLE

Solve $3x^2 - 6x + 2 = 0$ by 'completing the square'.

$$3x^2 - 6x + 2 = 0$$
$$\Rightarrow 3\left(x^2 - 2x + \frac{2}{3}\right) = 0$$
$$\Rightarrow 3\left((x - 1)^2 - \frac{1}{3}\right) = 0$$
$$\Rightarrow (x - 1)^2 = \frac{1}{3}$$
$$\Rightarrow x - 1 = \pm\frac{1}{\sqrt{3}}$$
$$\Rightarrow x = 1 \pm \frac{1}{\sqrt{3}}$$

METHOD NOTE

First, take a factor outside a bracket, so that what's left starts with x^2.

Then, choose an expression like $(x - k)^2$ so that the first 2 coefficients are correct. Here, we used $(x - 1)^2 = x^2 - 2x + 1$. The '-1' is half of the 'middle term'.

Finally, adjust the last term to make the equation correct.

EXAM NOTE

Don't forget the 'plus or minus'!

Use the quadratic formula to solve $3x^2 - 6x + 2 = 0$

The formula gives $x = \dfrac{-b \pm \sqrt{b^2 - 4ac}}{2a}$

$$= \dfrac{6 \pm \sqrt{(-6)^2 - 4 \times 3 \times 2}}{2 \times 3}$$

$$= \dfrac{6 \pm \sqrt{12}}{6} = 1 \pm \sqrt{\dfrac{12}{36}} = 1 \pm \dfrac{1}{\sqrt{3}}$$

as before.

Mathematicians often say 'two equal real roots' instead of 'one real root'. If you go on to study Complex Numbers you will see that the 'no real roots' case can be said to have 'two complex roots'. But you don't need to know about complex numbers for AS-Level.

If $b^2 - 4ac = 0$, then the quadratic expression will be a 'perfect square' and it will factorise to give $(px + q)^2$ for suitable p, q.

Because the formula includes the term $\sqrt{b^2 - 4ac}$, there can be no real solutions of a quadratic equation if this term, called the **discriminant**, is negative. Also if the discriminant is zero, there will be just one real solution, which means that the quadratic graph will just touch the x-axis.

So,

$b^2 - 4ac > 0 \Rightarrow$ equation has two real roots

$b^2 - 4ac = 0 \Rightarrow$ equation has one real root

$b^2 - 4ac < 0 \Rightarrow$ equation has no real roots

EXAMPLE

For what values of b does $5x^2 + bx + 1 = 0$ have no real solutions?

For no real roots, the discriminant is negative

$$b^2 - 4 \times 5 \times 1 < 0 \Rightarrow b^2 - 20 < 0$$

$$b^2 = 20 \Rightarrow b = \pm 2\sqrt{5}$$

so $\qquad b^2 - 20 < 0 \Rightarrow -2\sqrt{5} < b < +2\sqrt{5}$

Powers and indices

Practise manipulating indices and remember the rules.

$x^p \times x^q = x^{p+q}$ e.g. $x^3 \times x^4 = x^7$

$x^p \div x^q = x^{p-q}$ e.g. $x^7 \div x^2 = x^5$

$(x^p)^q = x^{p \times q}$ e.g. $(x^3)^4 = x^{12}$

$x^{-p} = \dfrac{1}{x^p}$ e.g. $x^{-2} = \dfrac{1}{x^2}$

$x^0 = 1$

$x^{\frac{1}{2}} = \sqrt{x}$, and $x^{\frac{1}{3}} = \sqrt[3]{x}$ and so on, as $x^{\frac{1}{2}} \times x^{\frac{1}{2}} = x^1 = x$

Polynomials

All of

$$x^2 - 3x + 1, \quad 3x^5 - 2x^4 + x - 2, \quad a + bx + cx^2 + dx^3$$

are called **polynomials**. Each term consists of a positive, integer power of x, multiplied by a constant, which could be negative or fractional.

All polynomials are functions, and many algebra, graph and calculus questions in AS-Level exams deal with polynomial functions.

The factor theorem

> For a polynomial $f(x)$, if $f(a) = 0$ for a constant a, then the term $(x - a)$ is a factor of $f(x)$.

This result is particularly useful when searching for **factors** of **polynomials**.

EXAMPLE

Factorise the polynomial

$$f(x) = x^3 - 3x^2 - 4x + 12$$

Using the factor theorem, it is usually best to start searching with simple integers, and search in a logical order.

Here,

$x = +1 \Rightarrow f(1) = 1 - 3 - 4 + 12 \neq 0,$

$x = -1 \Rightarrow f(-1) = -1 - 3 + 4 + 12 \neq 0,$

$x = +2 \Rightarrow f(2) = 8 - 12 - 8 + 12 = 0,$

and so $(x - 2)$ is a factor.
So factorising gives

$$x^3 - 3x^2 - 4x + 12 = (x - 2)(x^2 - x - 6)$$
$$= (x - 2)(x + 2)(x - 3)$$

REMEMBER

The statement that $y = f(x)$ is a function means that for any value of x, the mapping gives exactly one value of y.

EXAM NOTE

It's often easiest to try $f(x)$ for values of x: $+1$, -1, $+2$, -2, $+3$, etc.

You can try some simple fractions, like $\frac{1}{2}$, but in an exam, if simple whole numbers don't work you are likely to have made a mistake.

EXAMPLE

Both $(x - 1)$ and $(x + 1)$ are factors of

$$f(x) = 2x^3 + ax^2 + bx - 1$$

Find the values of the constants a and b.

Since $(x - 1)$ is a factor, then $f(1) = 0$, and since $(x + 1)$ is also a factor, then $f(-1) = 0$ also.

$$\left. \begin{array}{l} f(1) = 2 + a + b - 1 = 0 \Rightarrow a + b = -1 \\ f(-1) = -2 + a - b - 1 = 0 \Rightarrow a - b = 1 \end{array} \right\}$$

subtracting $\Rightarrow 2a = 2 \Rightarrow a = 1$

and substituting $\Rightarrow b = -2$

So $f(x) = 2x^3 + x^2 - 2x - 1$
$ = (x - 1)(x + 1)(2x + 1)$

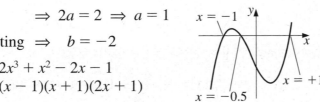

Substituting $f(1) = 0$ and $f(-1) = 0$ gives a pair of simultaneous equations in a and b.

Series

Definitions

A series is the sum of individual terms of a sequence

$$s_n = u_1 + u_2 + u_3 + \ldots + u_n$$

written as $s_n = \sum_{i=1}^{n} u_i$

where u_n represents the individual terms, and s_n is the sum of the first n terms.

If consecutive terms of the series are found by adding a constant value, the series is an **Arithmetic Progression** (called an 'AP'). The constant value is the **common difference**, d, and the first term is a.

For example

$$5 + 8 + 11 + 14 + \ldots \text{ is an AP with } a = 5, d = 3$$
$$6 + 2 - 2 - 6 - 10 - \ldots \text{ is an AP with } a = 6, d = -4$$

If consecutive terms of a series are found by multiplying by a constant factor, the series is a **Geometric Progression** (called a 'GP'). The constant factor is the **common ratio**, r, and the first term is a.

Arithmetic progressions

Using the notation above,

$$u_n = a + (n-1)d,$$
$$\text{also } u_n = u_{n-1} + d$$

$$s_n = \tfrac{1}{2}n\{2a + (n-1)d\}$$

The sum can also be written

$$s_n = \tfrac{1}{2}n(a + L) \text{ where } L \text{ is the last term.}$$

EXAMPLE

Find the sum of the series $6 + 15 + 24 + \ldots + 690$

Work out what you already know.

The series is an AP, with first term $a = 6$, and common difference $d = 9$. Using the fact that the last term is 690,

Now use the facts you have to find n, the only standard term you don't know. Here you can use u_n, the nth term, to find n.

$$u_n = a + (n-1)d \Rightarrow 6 + (n-1)(9) = 690$$
$$\Rightarrow \quad (n-1)(9) = 684$$
$$\Rightarrow \quad 9n - 9 = 684$$
$$\Rightarrow \quad 9n = 693$$
$$\Rightarrow \quad n = 77$$

So the sum of the series of 77 terms is given by

Now use the formula for s_n.

$$s_n = \tfrac{1}{2} \times 77[2 \times 6 + (77 - 1) \times 9] = 26\,796$$

Geometric progressions

If u_i is the ith term, s_n is the sum of the first n terms, a is the first term and r is the common ratio,

$$u_n = ar^{n-1}$$

also $\ u_n = ru_{n-1}$

$$s_n = \frac{a(r^n - 1)}{r - 1} \ \text{ or } \ \frac{a(1 - r^n)}{1 - r} \ \text{ and if } \ r < 1, \ s_\infty = \frac{a}{1 - r}$$

EXAMPLE

The second term of a GP is 7.5, and the fifth term of the GP is 25.3125.

Find the sum of the first ten terms.

You know that $u_2 = ar = 7.5$

and $\qquad\qquad u_5 = ar^4 = 25.3125$

so, by dividing, you can find the common ratio r

$$\frac{u_5}{u_2} = \frac{ar^4}{ar} = \frac{25.3125}{7.5}$$

$$\Rightarrow \ r^3 = 3.375$$

$$\Rightarrow \ \ r = 1.5$$

And from the second term,

$$a(1.5) = 7.5$$

$$\Rightarrow \ \ a = 5$$

So the sum of the first 10 terms

$$s_{10} = \frac{5(1.5^{10} - 1)}{1.5 - 1}$$

$$= \frac{5(57.665 - 1)}{0.5}$$

$$= 566.650 \ \text{(to 3 d.p.)}$$

EXAMPLE

Julie invests £500 in a savings account at the start of each year. At the end of each year, interest of 6% is added by the bank. How much does she have after 10 years?

Adding 6% interest at the end of the year is equivalent to multiplying the value of Julie's savings by 1.06.

So, after one year, her savings are £500 × 1.06

After two years they are £$(500 \times 1.06) + (500 \times 1.06^2)$

So, after 10 years they are

$$£(500 \times 1.06 + 500 \times 1.06^2 + \ldots + 500 \times 1.06^{10})$$

So her total savings are given by a GP, with first term $a = 500 \times 1.06 = 530$ and common ratio $r = 1.06$

The total is $\ s_{10} = \dfrac{530(1.06^{10} - 1)}{1.06 - 1} = 6985.82 \ \text{(to 2 d.p.)}$

Julie has £6985.82 after 10 years.

Algebra and series

1 For the equation

$$3x^2 + px + 4 = 0$$

find the range of values of p such that the equation has

(*a*) no real roots

(*b*) one real root

(*c*) two real roots.

2 Solve, by completing the square,

$$4x^2 - 2x + 5 = 0$$

leaving your answer in surd form.

3 The expression $(x - 2)$ is a factor of the polynomial

$$p(x) = 2x^3 - x^2 + kx - 16$$

(*a*) Find the value of k.

(*b*) Prove that there are no other real roots of the equation

$$p(x) = 0$$

4 (*a*) The sum of this series is 610.

$$2 + 5 + 8 + 11 + \dots = 610$$

How many terms are there in the series?

(*b*) Find the initial term and the common ratio of the geometric progression for which

$$u_3 = 4 \quad \text{and} \quad u_6 = \tfrac{1}{2}$$

(*c*) For the geometric progression in (*b*)
   (i) show that the sum to infinity is 32
   (ii) find the number of terms necessary for the sum to exceed 31.99.

5 Jane is offered a choice of two pay schemes when she starts her new job.

   Scheme I: £12 000 in year 1, and then an increase of £500 at the start of each year.

   Scheme II: £12 000 in year 1, and then an increase of 4% at the start of each year.

(*a*) By using formulae for AP's and GP's, find how much Jane will earn in year 4 on each scheme, correct to the nearest pound.

(*b*) By using formulae for AP's and GP's, find how much Jane will earn in total in 10 years, correct to the nearest pound.

4 Trigonometry

Trigonometric functions

The basic trigonometric functions **sin**, **cos** and **tan** are shown in the graph.

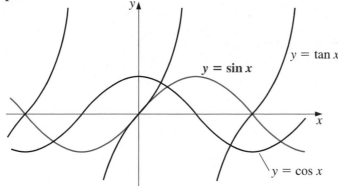

The graphs of sine and cosine have exactly the same shape, but sine passes through the origin, while cosine cuts the x-axis at $\pm 90°$ or $\pm \dfrac{\pi}{2}$ radians.

> All three graphs are periodic. sin and tan are odd functions and cos is an even function.

Since the cos graph can be found from the sin graph by a translation to the left of $90°$ or $\dfrac{\pi}{2}$ radians, then

> $\cos x° \equiv \sin(x° + 90°)$ or $\cos x \equiv \sin\left(x + \dfrac{\pi}{2}\right)$ in radians

Definitions

> $$\tan x \equiv \frac{\sin x}{\cos x}$$
> and also $\sec x \equiv \dfrac{1}{\cos x}$ $\cosec x \equiv \dfrac{1}{\sin x}$ $\cot x \equiv \dfrac{1}{\tan x}$

Useful results:

> $\sin^2 \theta + \cos^2 \theta \equiv 1$
> $1 + \tan^2 \theta \equiv 1$
> $\cot^2 \theta + 1 \equiv \cosec^2 \theta$

EXAMPLE

Find all the solutions of $1 + 2\sin 2x = 0$ for values of x such that $-180° < x° < +180°$.

$$1 + 2\sin 2x = 0 \Rightarrow \sin 2x = -\tfrac{1}{2}$$

so $2x = -150°, -30°, +210°, +330°$

and $x = -75°, -15°, +105°, +165°$

Notice that sin x crosses the x-axis at $x = -360°, -180°, 0°, +180°, +360°$ while cos x crosses at $x = -270°, -90°, +90°, +270°$. For the graphs of sin x and cos x, $-1 \leqslant f(x) \leqslant +1$

REVISION NOTE

It's well worth knowing some exact values.

	sin	cos	tan	rad
0°	0	1	0	0
30°	$\dfrac{1}{2}$	$\dfrac{\sqrt{3}}{2}$	$\dfrac{1}{\sqrt{3}}$	$\dfrac{\pi}{6}$
45°	$\dfrac{1}{\sqrt{2}}$	$\dfrac{1}{\sqrt{2}}$	1	$\dfrac{\pi}{4}$
60°	$\dfrac{\sqrt{3}}{2}$	$\dfrac{1}{2}$	$\sqrt{3}$	$\dfrac{\pi}{3}$
90°	1	0	∞	$\dfrac{\pi}{2}$

METHOD NOTE

The symbol $\equiv$ means **'identically equal'**. It means these expressions are true for *all* values of x.

REVISION NOTE

You need to know these definitions.

METHOD NOTE

Because we want values of x between $-180°$ and $+180°$, we look for all values of $2x$ between $-360°$ and $+360°$.

Your calculator will give arcsin (-0.5) or $\sin^{-1}(-0.5)$ as $-30°$, in 'degrees' mode.

From a graph, $\sin \alpha$ is negative in the third and fourth quadrants.

Inverse trigonometric functions

$$\sin 30° = \sin 150° = 0.5$$

So the function $f(x) = \sin x$ is a many-to-one function.

To make the inverse mapping a function, for

$$y = \sin^{-1} x$$

working in radians, the domain is $-1 \leqslant x \leqslant +1$ but the range of the function is restricted to $-\dfrac{\pi}{2} \leqslant y \leqslant +\dfrac{\pi}{2}$.

The inverse mapping could be written as 'the angle which has sine equal to 0.5'. The outcome could be 30°, or 150°, or other answers.

HINT

If you are not clear about radians, look at page 35 first.

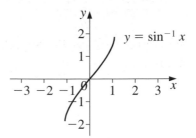

The graph of $y = \sin^{-1} x$ above shows the restricted domain. There are no values for $x < -1$, or for $x > +1$, and the graph extends only from

$$y = -\dfrac{\pi}{2} \quad \text{to} \quad y = +\dfrac{\pi}{2}.$$

REVISION NOTE

You will need $\sin^{-1} x$ and $\tan^{-1} x$ in some calculus questions. Make sure you know how your calculator handles **inverse trig. functions**.

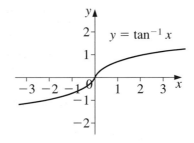

$y = \tan^{-1} x$ does not have a restricted domain. Any value of x is possible, but the only possible values of the range are

$$-\dfrac{\pi}{2} < y < +\dfrac{\pi}{2}.$$

METHOD NOTE

You should recognise that $\sin\dfrac{\pi}{3} = \dfrac{\sqrt{3}}{2}$. Your calculator will give you just one result, because of the restricted range of $\sin^{-1} x$, but you need to find the other values. Here, because $-2\pi < x \leqslant +2\pi$ you need all the values for

$$-\pi < \tfrac{1}{2}x \leqslant +\pi$$

EXAMPLE

Solve the equation $2 \sin \tfrac{1}{2}x = \sqrt{3}$ for values of x such that $-2\pi < x \leqslant +2\pi$

$$2 \sin \tfrac{1}{2} x = \sqrt{3}$$

$$\Rightarrow \quad \sin \tfrac{1}{2} x = \dfrac{\sqrt{3}}{2}$$

$$= \sin \dfrac{\pi}{3}$$

$$\Rightarrow \quad \tfrac{1}{2}x = \dfrac{\pi}{3}, \dfrac{2\pi}{3}$$

$$\Rightarrow \quad x = \dfrac{2\pi}{3}, \dfrac{4\pi}{3}$$

The sine rule and the cosine rule

You may have met the sine and cosine rules in your GCSE course, but they are extremely useful at AS-Level, especially to help 'solve' triangles in Mechanics questions.

The sine rule

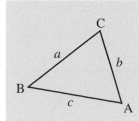

In any triangle

$$\frac{\sin A}{a} = \frac{\sin B}{b} = \frac{\sin C}{c}$$

where A, B, C are the angles of the triangle, and a, b, c are the opposite sides.

REVISION NOTE

This applies to any triangle … right-angled, isosceles, scalene, whatever.

EXAMPLE

In the triangle, find the length b, and the angle at A.

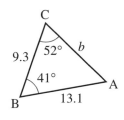

Using the sine rule

$$\frac{b}{\sin 41°} = \frac{13.1}{\sin 52°}$$

$$\Rightarrow \qquad b = \frac{13.1 \times \sin 41°}{\sin 52°}$$

$$= \frac{(13.1)\,(0.6561)}{0.7880}$$

$$= 10.91 \text{ (to 2 d.p.)}$$

$$\text{Angle A} = 180° - 52° - 41°$$

$$= 87°$$

EXAM NOTE

Don't find a 'hard' way to do this. In AS-Level, the three angles of a triangle *still* add up to 180°.

The 'ambiguous case'

In this triangle you can use the sine rule to find angle B.

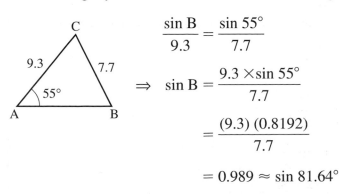

$$\frac{\sin B}{9.3} = \frac{\sin 55°}{7.7}$$

$$\Rightarrow \quad \sin B = \frac{9.3 \times \sin 55°}{7.7}$$

$$= \frac{(9.3)\,(0.8192)}{7.7}$$

$$= 0.989 \approx \sin 81.64°$$

So angle B is 81.6°, but also sin 98.3° = 0.989, so there is another possible answer. The diagram suggests B is less than a right angle, but if the diagram is not to scale you need more information to know which is correct.

The triangle could be shaped:

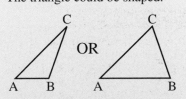

OR

Angle B is close to 90°, so you need to know more before you can decide.

NOTE

In any triangle, the largest angle must be opposite the longest side. The smallest angle is opposite the shortest side.

The cosine rule

REVISION NOTE

You can write down two equivalent formulae to give you values for a^2 and b^2.

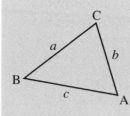

In any triangle

$$c^2 = a^2 + b^2 - 2ab \cos C$$

where A, B, C are the angles of the triangle, and a, b, c are the opposite sides.

EXAMPLE

In the triangle, find:

(a) the length of the third side.

(b) the remaining two angles.

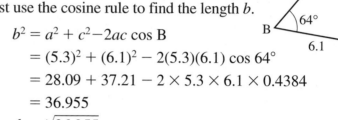

First use the cosine rule to find the length b.

$$b^2 = a^2 + c^2 - 2ac \cos B$$
$$= (5.3)^2 + (6.1)^2 - 2(5.3)(6.1) \cos 64°$$
$$= 28.09 + 37.21 - 2 \times 5.3 \times 6.1 \times 0.4384$$
$$= 36.955$$
$$\Rightarrow \quad b = \sqrt{36.955}$$
$$= 6.08 \text{ (to 2 d.p.)}$$

Now use the sine rule to find angle A.

$$\frac{\sin A}{5.3} = \frac{\sin 64°}{6.08}$$

$$\Rightarrow \quad \sin A = \frac{5.3 \sin 64°}{6.08}$$

$$= \frac{5.3 \times 0.8988}{6.08} = 0.7835 \approx \sin 51.6°$$

so A is 51.6°, and $C = 180° - 64° - 51.6° = 64.4°$.

EXAM NOTE

As with the sine rule example, don't use some complicated method to find the third angle.

EXAMPLE

In this triangle, find angle C.
Describe the triangle.

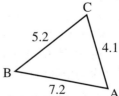

Using the cosine rule,

$$(7.2)^2 = (5.2)^2 + (4.1)^2 - 2(5.2)(4.1) \cos C$$

$$\Rightarrow \quad \cos C = \frac{(5.2)^2 + (4.1)^2 - (7.2)^2}{2(5.2)(4.1)}$$

$$= -0.1874 = \cos 100.8° \text{ (to 1 d.p.)}$$

So C is 100.8°, and ABC is an obtuse scalene triangle.

EXAM NOTE

There is no **ambiguous case** for the cosine rule.

$\cos x°$ is positive for $0° < x° < 90°$ but negative for $90° < x° < 180°$

Radians

Although degrees are used to measure angles for all everyday purposes, and are generally used in geometrical diagrams in AS-level, the alternative measure called the **radian** is extremely important in mathematics.

The radian is defined geometrically below, but radian measure does crop up 'naturally' in applying calculus to trigonometric functions.

REVISION NOTE

You don't need to reproduce the calculus definition of a radian for AS-Level.

Definition

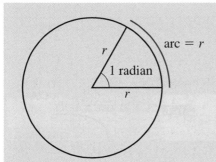

A radian is the angle subtended at the centre of a circle by an arc of length equal to the radius.

Important results are:

2π radians $= 360°$

π radians $= 180°$

The circumference of a circle is $2\pi r$, so $360° = 2\pi$ in radians. So 1 radian $\approx 57.3°$

Arc length and area of a sector

Using radians as a measure of angle, there are two straightforward results.

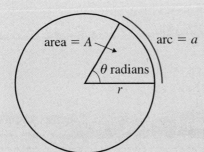

The arc subtended by an angle θ radians at the centre of the circle, and the area enclosed in the sector with angle θ radians at the centre are:

Arc length $\quad a = r\theta$

Area of sector $A = \frac{1}{2}r^2\theta$

EXAMPLE

The 'deep' section of a circular swimming pool of radius 18 m is in the form of a sector of the circle. The circular edge of the deep section has length 40 m. Find the area of the surface of the pool above the deep section.

If the angle subtended at the centre of the pool by the deep section is θ, in radians

$$a = r\theta \implies 40 = (18)\theta$$
$$\implies \theta = 2.22\ldots$$

and so $\quad A = \frac{1}{2}r^2\theta \implies A = \frac{1}{2}(18)^2(2.22\ldots)$

$$= 360 \text{ m}^2$$

EXAM NOTE

Don't forget to put your calculator into 'radian mode'. Get into the habit of checking.

Exam boards have different ideas about this topic. For some boards these results are only in A2, not AS. Check for your Board.

NOTE

Here, the $\pm$ and $\mp$ mean 'plus or minus'. The top symbols correspond in each formula, and the bottom ones.

EXAM NOTE

Because the question says 'exact', don't put decimals in here. Leave it as this, or $\dfrac{\sqrt{2}(1 + \sqrt{3})}{4}$ if you prefer.

EXAM NOTE

You might be asked to find one of these formulae, but using the results is not really in AS-Level.

Just put A = B in one of the compound formulae. It's worth practising.

NOTE

Degrees are not mentioned – the question is in radians.

From $\dfrac{r \sin \alpha}{r \cos \alpha} = \dfrac{4}{3}$ by dividing.
By squaring and adding
$r^2 \cos^2 \alpha + r^2 \sin^2 \alpha = 3^2 + 4^2$
Use $\sin^2 \alpha + \cos^2 \alpha = 1$

REMEMBER

Check that these are all the solutions in $-\pi < x < +\pi$

Trigonometric formulae

These formulae are helpful in finding exact values, simplifying trig. equations, or transforming integrals with trig. functions.

Compound angles

$$\sin (A \pm B) = \sin A \cos B \pm \cos A \sin B$$
$$\cos (A \pm B) = \cos A \cos B \mp \sin A \sin B$$
$$\tan (A \pm B) = \frac{\tan A \pm \tan B}{1 \mp \tan A \tan B}$$

EXAMPLE

Find an exact value for $\sin 75°$.
$$\begin{aligned}
\sin 75° &= \sin (45° + 30°) \\
&= \sin 45° \cos 30° + \cos 45° \sin 30° \\
&= \frac{1}{\sqrt{2}} \cdot \frac{\sqrt{3}}{2} + \frac{1}{\sqrt{2}} \cdot \frac{1}{2} \\
&= \frac{1 + \sqrt{3}}{2\sqrt{2}}
\end{aligned}$$

Double angles

$$\begin{aligned}
\sin 2A &= 2 \sin A \cos A \\
\cos 2A &= \cos^2 A - \sin^2 A \\
&= 2 \cos^2 A - 1 \\
&= 1 - 2 \sin^2 A \\
\tan 2A &= \frac{2 \tan A}{1 - \tan^2 A}
\end{aligned}$$

The expression $a \cos x + b \sin x$

By comparing the expression $a \cos x + b \sin x$ with the right-hand side of the compound angle formula for $\cos (A + B)$, it is possible to write it as $r \cos (x + \alpha)$.

EXAMPLE

Solve $3 \cos x + 4 \sin x = 2$ for $-\pi < x < +\pi$

Comparing $r \cos x \cos \alpha + r \sin x \sin \alpha = r \cos (x - \alpha)$ with the l.h.s. of the equation, $r \cos \alpha = 3$ and $r \sin \alpha = 4$ because, for example, the coefficient of $\cos x$ is $r \cos \alpha$ in the expansion for $r \cos (x - \alpha)$.

So if $r \cos \alpha = 3$ and $r \sin \alpha = 4$, then we can get
$$\tan \alpha = \tfrac{4}{3} \text{ and } r^2 = 3^2 + 4^2 = 25$$
$$\Rightarrow \quad \alpha = 0.9237 \text{ (rads) and } r = 5$$
So $\quad 5 \cos (x - 0.9237) = 2$
$$\Rightarrow \quad \cos (x - 0.9237) = \tfrac{2}{5} = \cos (1.159) \text{ (rads)}$$
$$\Rightarrow \quad x - 0.9237 = 1.159 \text{ or } -1.159, \text{ so } x = 2.08 \text{ or } -0.24$$

Trigonometry

1 Solve these equations for all values of x such that $-180° < x° \leqslant +180°$, giving x to one decimal place.

(a) $2 + 3 \cos x = 0$

(b) $3 + 4 \sin 3x = 0$

2 A crewman in a small boat takes a bearing on a steeple on the shoreline, and finds that the bearing of the steeple is 063°. The boat then sails 2 kilometres due North, and the crewman finds that the bearing of the steeple is now 106°.

By drawing a suitable triangle, and calculating, find the distance to the steeple at each sighting.

3 The field in the diagram is being surveyed. Each hedge is straight, and the angle between the edges AB and AD is 74°. The lengths of the hedges AB and AD are 62 metres and 47 metres. Find the length of the straight footpath which crosses the field from B to D.

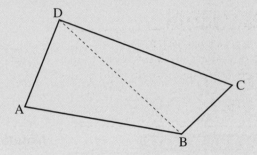

4 A circular raceway on ice has a radius of 40 metres at the inner edge of lane one. The inner edge of lane two has a radius of 44 metres.

What distance 'ahead' of lane one should the lane two starting position be marked, if both are marked on the inside edge, and the starting points are marked for a 'fair' race of one lap?

5 Solve the equation

$$5 \sin x - 12 \cos x = 10$$

for $-\pi < x \leqslant +\pi$.

6 (a) Use the identity

$$\sin (A - B) = \sin A \cos B - \cos A \sin B$$

with $A = 45°$ and $B = 30°$, to find an exact value for sin 15°.

(b) By putting $A = B$ in the identity

$$\sin (A + B) = \sin A \cos B + \cos A \sin B$$

find an identity for sin 2A.

7 A child's kite has a central stick 50 cm long. The shorter edges of the kite are each 25 cm long. The longer edges are each 40 cm.

Find the angles at the vertices of the kite.

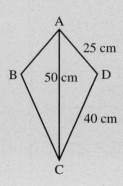

5 Calculus – understanding change

Gradient of a curve

The gradient of a curve at any point is given by the gradient of its tangent at that point.

For a curve given by a function $y = f(x)$, the process of finding the gradient of the tangent is **differentiation**.

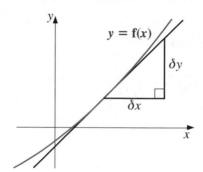

In the diagram, the gradient of the tangent is given by

$$\lim\limits_{\delta x \to 0} \frac{\delta y}{\delta x}$$

If we do this process formally, for the function $y = f(x)$ we get the derivative $\dfrac{dy}{dx}$ and by substituting a value of x, we can find the gradient of the tangent at a particular point.

Notation

There are two standard notations in use in AS-Level, and you need to be comfortable with both.

The derivative of a function $y = f(x)$ is written either as $\dfrac{dy}{dx}$ or as $f'(x)$.

Similarly there are two notations for second derivatives and higher derivatives:

$$y = f(x)$$
$$\Rightarrow \quad \frac{dy}{dx} = f'(x)$$
$$\Rightarrow \quad \frac{d^2y}{dx^2} = f''(x)$$

… and so on.

Basic differentiation results

You need to learn the basic differentiation results:

For $f(x) = \ldots$	You get $f'(x) = \ldots$
x^n	nx^{n-1}
$\sin x$	$\cos x$
$\cos x$	$-\sin x$
e^x	e^x
$\ln x$ (or $\log_e x$)	$\dfrac{1}{x}$

Differentiating algebraic expressions

The table on page 38 gives the basic results to remember for differentiating algebraic expressions. The following two rules help with expressions made from simpler functions.

Addition of two functions

If $y = f(x) + g(x)$ then $\dfrac{dy}{dx} = f'(x) + g'(x)$

Multiplication of a function by a constant

If $y = k[f(x)]$ with k constant, then $\dfrac{dy}{dx} = k[f'(x)]$

REVISION NOTE

These results do not help when two functions are multiplied together, or when one is divided by another.

These cases, and some others, are covered in Unit 6.

These two results mean that differentiation is a linear operation. *You don't need to know that for AS-Level*, but it is very important to mathematicians.

Dealing with fractional powers

The rule $y = x^n \Rightarrow \dfrac{dy}{dx} = nx^{n-1}$ can be applied directly to deal with fractional powers of x.

EXAMPLE

Find the derivatives of (a) $\sqrt[3]{x}$ and (b) $\dfrac{1}{\sqrt{x}}$

(a) $f(x) = \sqrt[3]{x} = x^{\frac{1}{3}} \Rightarrow f'(x) = \frac{1}{3}x^{-\frac{2}{3}}$

(b) $y = \dfrac{1}{\sqrt{x}} = x^{-\frac{1}{2}} \Rightarrow \dfrac{dy}{dx} = -\frac{1}{2}x^{-\frac{3}{2}} = -\dfrac{1}{2\sqrt{x^3}}$

REVISION NOTE

Notice the two different notations used for the two parts of the question. Make sure you can use both.

Using all the results

EXAMPLE

Differentiate (a) $(x^3 - 2)x^2$ (b) $\sqrt{x} + \dfrac{4}{x^3} - 3x$

(a) $y = (x^3 - 2)x^2 \Rightarrow y = x^5 - 2x^2$

$\Rightarrow \dfrac{dy}{dx} = 5x^4 - 2(2x)$

$= x(5x^3 - 4)$

(b) $f(x) = \sqrt{x} + \dfrac{4}{x^3} - 3x \Rightarrow f(x) = x^{\frac{1}{2}} + 4x^{-3} - 3x$

$\Rightarrow f'(x) = \frac{1}{2}x^{-\frac{1}{2}} + 4(-3)x^{-4} - 3$

$= -\dfrac{1}{2\sqrt{x}} - \dfrac{12}{x^4} - 3$

METHOD NOTE

You need to multiply out the brackets here. See Unit 6 for another method.

Remember to 'tidy up' the answer. Here, 'x' is a factor in both terms.

Write each term in the form x^n then differentiate to nx^{n-1} and tidy up as much as possible.

Always try to simplify your answers. Here, it is easy in part (a), but part (b) is not worth manipulating further.

Finding the tangent to a graph

Many questions at AS-Level require you to find the equation of a tangent to a graph. The examples below use differentiation to find the gradient, and the equation of a straight line

$$y = mx + c$$

to give the tangent.

EXAMPLE

For the graph $y = x^3 - x^2 - 5x + 2$, find the equation of the tangent at $x = 2$, and the area enclosed between the x-axis, the y-axis and that tangent.

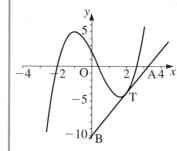

First sketch the graph (see Unit 1).

At the point T,

$$x = 2 \Rightarrow y = 8 - 4 - 10 + 2 = -4$$

To find the equation of the tangent at T $(2, -4)$

$$y = x^3 - x^2 - 5x + 2$$

$$\Rightarrow \frac{dy}{dx} = 3x^2 - 2x - 5$$

At T, $x = 2 \Rightarrow \frac{dy}{dx} = 3 \times 2^2 - 2 \times 2 - 5 = 3$

So the equation of the tangent is $y = 3x + c$, and it passes through T, with coordinates $(2, -4)$.

Hence

$$-4 = 3 \times 2 + c$$

$$\Rightarrow \quad c = -10$$

So the tangent is $y = 3x - 10$.

If the tangent cuts the axes at A and B, as shown in the diagram, then A is $(\frac{10}{3}, 0)$ and B is $(0, -10)$.

So the area between the tangent and the axes, which is triangle OAB in the diagram, is

$$\text{area} = \tfrac{1}{2} \times \tfrac{10}{3} \times 10 = \tfrac{50}{3} = 16\tfrac{2}{3} \text{ units}^2$$

EXAMPLE

In the example above, for what other value of x is the tangent to the graph parallel to the tangent at the point T?

Because the tangent at T has gradient 3, we can find another point where the derivative is 3.

$$\frac{dy}{dx} = 3 \Rightarrow 3x^2 - 2x - 5 = 3$$

$$\Rightarrow \quad 3x^2 - 2x - 8 = 0$$

$$\Rightarrow \quad (3x + 4)(x - 2) = 0$$

$$\Rightarrow \quad x = -\tfrac{4}{3} \quad \text{or} \quad x = 2$$

and so for $x = -\tfrac{4}{3}$ the graph and its tangent are parallel to the tangent at the point T.

Finding a normal to a curve

At any point, the normal to a curve is the straight line through the point that is perpendicular to the tangent.

Using a standard result about gradients:

If the gradient of the tangent is m_1, and the gradient of the normal is m_2, then $m_1 m_2 = -1$.

EXAMPLE

Sketch the graph $y = x^3 - 2x^2 + 3$. At what other points does the normal at the point where $x = 1$ cross the graph?

First, draw a sketch of the graph (see Unit 1).

Sketch the normal to the graph at $x = 1$. It also crosses the graph at A and B.

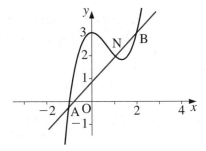

From the sketch graph it is not obvious whether the crossing point A is above or below the x-axis.

Next, find the gradient of the graph at N, where $x = 1$.

$$x = 1 \quad \Rightarrow \quad y = 1^3 - 2 \times 1^2 + 3 = 2 \text{ so point N is } (1, 2).$$

Now
$$y = x^3 - 2x^2 + 3$$
$$\Rightarrow \quad \frac{dy}{dx} = 3x^2 - 4x$$
$$x = 1 \quad \Rightarrow \quad \frac{dy}{dx} = 3 \times 1^2 - 4 \times 1 = -1$$

So the gradient of the curve (and also of the tangent) is -1.

So, if the gradient of the normal is m_2.

$$(-1)m_2 = -1 \quad \Rightarrow \quad m_2 = +1$$

and the normal is $y = (+1)x + c$.

Substituting the coordinates of N

$$2 = 1 + c$$
$$\Rightarrow \quad c = 1$$

The normal at $x = 1$ is $y = x + 1$

To find the points where the normal and the curve cross, solve the equations for the line and the curve simultaneously.

$$\left. \begin{array}{l} y = x^3 - 2x^2 + 3 \\ y = x + 1 \end{array} \right\}$$

$$\Rightarrow \quad (x + 1) = x^3 - 2x^2 + 3$$
$$\Rightarrow \quad x^3 - 2x^2 - x + 2 = 0$$
$$\Rightarrow \quad (x - 1)(x^2 - x - 2) = 0$$
$$\Rightarrow \quad (x - 1)(x - 2)(x + 1) = 0$$
$$\Rightarrow \quad x = 1 \text{ or } x = 2 \text{ or } x = -1$$
and $y = 2$ or $y = 3$ or $y = 0$

So the normal to the curve at $x = 1$ also crosses the curve at the points A $(-1, 0)$ and B $(2, 3)$.

EXAM NOTE

It's easy to forget to do this. Don't find the equation of the tangent by mistake.

Find m, then find c by substituting the known coordinates into the equation.

METHOD NOTE

Just put the two '$y = \ldots$' expressions equal to each other and simplify.

Now factorise. You already know that $x - 1$ is a factor, because the normal crosses the curve when $x = 1$.

Substituting the x values in the equation of the normal to find *both* coordinates of the points.

Maxima and minima

Differentiation gives a method of finding maximum or minimum points on a curve.

Because the tangent to the curve at a maximum or minimum point is horizontal (parallel to the x-axis), the gradient of the curve is zero. Such points on a graph are called **stationary points**.

gradient = 0

At a maximum or minimum point on a curve,
$$\frac{dy}{dx} = 0$$

Maximum and minimum points can be identified in most cases by using the second derivative. The diagram below shows the sign of the gradient at different points on the graph:

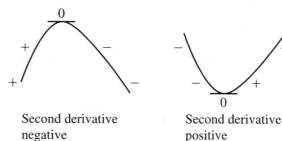

Second derivative negative

Second derivative positive

This change in sign of the gradient, $\dfrac{dy}{dx}$, can be detected by looking at the sign of the second derivative at the maximum or minimum.

Basic curve shapes

The results above can be summarised to give the basic shape of the curve for values of $\dfrac{dy}{dx}$.

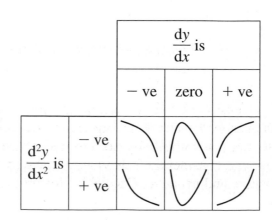

		$\dfrac{dy}{dx}$ is		
		− ve	zero	+ ve
$\dfrac{d^2y}{dx^2}$ is	− ve			
	+ ve			

Maxima and minima (cont.)

EXAMPLE

Find the stationary points on the graph

$$y = 2x^3 - 3x^2 - 12x - 4$$

Differentiating the equation twice,

$$y = 2x^3 - 3x^2 - 12x - 4$$

$$\Rightarrow \quad \frac{dy}{dx} = 6x^2 - 6x - 12$$

$$\Rightarrow \quad \frac{d^2y}{dx^2} = 12x - 6$$

So for the stationary points

$$\frac{dy}{dx} = 0 \Rightarrow 6(x + 1)(x - 2) = 0$$

$$\Rightarrow \quad x = -1 \text{ or} \quad x = \quad 2$$

$$\text{and } y = \quad 3 \text{ or} \quad y = -24$$

and also $\quad \dfrac{d^2y}{dx^2} < \quad 0 \text{ or } \dfrac{d^2y}{dx^2} > \quad 0$

So $(-1, 3)$ is a maximum point, and $(2, -24)$ is a minimum point.

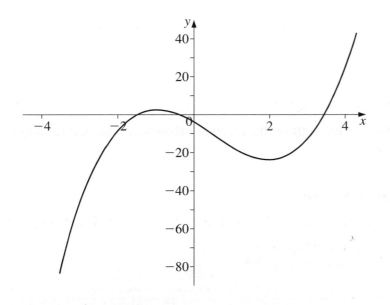

A sketch of the graph shows the two stationary points.

NOTE

This example has a cubic polynomial function. Exactly the same methods can be applied to more complicated functions.

METHOD NOTE

Put $\dfrac{dy}{dx} = 0$, and factorise.

Check the sign of $\dfrac{d^2y}{dx^2}$ at the stationary points.

In an exam, questions about finding stationary points are often combined with questions about sketching graphs. In order to complete the sketch you will usually need to find where the curve crosses the axes, and consider the behaviour of the curve for large values of x, positive and negative. Check also for prohibited values.

REVISION NOTE

All of this material is covered in Unit 1.

Points of inflexion

There is another possible stationary point, where the tangent is horizontal, and the gradient of the curve at the stationary point is equal to zero.

On each of the graphs below, there is a point in the middle of the 'S-bend' where the gradient of the graph is zero, and the tangent is horizontal and crosses the graph. These points are marked as A and B.

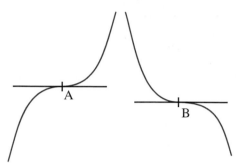

> A stationary point where the gradient of the graph is zero and the gradient of the tangent is also zero, is called a **point of inflexion**.

The tangents to the graphs at the points A and B are horizontal, so these points also satisfy

$$\frac{dy}{dx} = 0$$

Also, the sign of the gradient does not change from one side of the point of inflexion to the other, so at the points of inflexion

$$\frac{d^2y}{dx^2} = 0$$

Unfortunately, there are also special cases where $\frac{d^2y}{dx^2} = 0$ at either a maximum or minimum point, so:

> The only simple test for a point of inflexion is to check that the gradient does not change sign from one side of the stationary point to the other.

So, at a stationary point with $\frac{dy}{dx} = 0$

if $\frac{d^2y}{dx^2}$ is…	negative	the point is a maximum
	zero	the point could be a maximum, a minimum, or a point of inflexion. Check the sign of $\frac{dy}{dx}$ each side of the point.
	positive	the point is a minimum

METHOD NOTE

You can also have points of inflexion where the gradient is not horizontal, but there is an 'S-bend', and the curve crosses its own tangent. Such points are easy to find, because $\frac{d^2y}{dx^2} = 0$ but $\frac{dy}{dx} \neq 0$.

For example, the graph $y = x^4$ has a minimum at the origin, but it is easy to check that $\frac{d^2y}{dx^2} = 0$ there.

REVISION NOTE

For most AS-Levels, you will only have to identify 'horizontal' points of inflexion. Check your specification.

Points of inflexion (cont.)

EXAMPLE

Find the stationary points on the graph of

$$y = x^3 - 6x^2 + 3kx - 2$$

when (a) $k = 3$, and (b) $k = 4$.

(a) $k = 3$. Differentiating twice to find the stationary points,

$$y = x^3 - 6x^2 + 9x - 2$$

$$\Rightarrow \quad \frac{dy}{dx} = 3x^2 - 12x + 9$$

$$\Rightarrow \quad \frac{d^2y}{dx^2} = 6x - 12$$

$$\frac{dy}{dx} = 0 \Rightarrow 3(x - 1)(x - 3) = 0$$

$$\Rightarrow \quad x = 1 \text{ or } x = 3$$

$$\text{and} \quad y = 2 \text{ or } y = -2$$

$$\text{when} \quad \frac{d^2y}{dx^2} < 0 \text{ or } \frac{d^2y}{dx^2} > 0$$

and so $(1, 2)$ is a maximum, and $(3, -2)$ is a minimum.

(b) $k = 4$. The same method gives

$$y = x^3 - 6x^2 + 12x - 2$$

$$\Rightarrow \quad \frac{dy}{dx} = 3x^2 - 12x + 12$$

$$\Rightarrow \quad \frac{d^2y}{dx^2} = 6x - 12$$

$$\frac{dy}{dx} = 0 \Rightarrow 3(x - 2)^2 = 0$$

$$\Rightarrow x = 2 \text{ (twice)}$$

$$\text{and } y = 6$$

In this case, $\frac{d^2y}{dx^2} = 0$, so the stationary point $(2, 6)$ could be any of a minimum, a maximum, or a point of inflexion. Testing the sign of the gradient close to, but on either side of, the stationary point,

$$\text{at } x = 1.9, \quad \frac{dy}{dx} = +0.03$$

$$\text{and at } x = 2.1, \quad \frac{dy}{dx} = +0.03 \text{ also,}$$

so the sign of the gradient does not change, and $(2, 6)$ is a point of inflexion.

METHOD NOTE

Solve for $\frac{dy}{dx} = 0$, and find the corresponding values of x, y and $\frac{d^2y}{dx^2}$ to find the different stationary points.

REVISION NOTE

Using a graphical calculator, the graph looks like this:

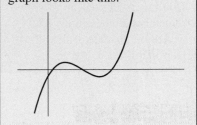

NOTE

Notice that $x = 2$ is a root of $\frac{dy}{dx} = 0$ twice. This suggests that two stationary points have been 'squashed together'.

REVISION NOTE

Using a graphical calculator, the graph looks like this:

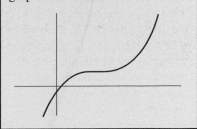

Integration

The process that reverses differentiation is called **integration**.

NOTATION

Just as with differentiation, the notation changes depending on the variables involved. If you start with $\dfrac{ds}{dt}$ then the integral will be $s = \int \ldots dt$

Notation

If $\dfrac{dy}{dx} = f(x)$ then $y = \int f(x)\, dx$.

The expression $\int f(x)\, dx$ is read as 'the integral of f(x) with respect to x'.

However, when you differentiate any of $x^3 + 1$, $x^3 + 4$, $x^3 - 2$, ..., or any other function of the form $x^3 +$ constant, you get the same result: $3x^2$. So whenever you carry out an integration you must remember to add a constant. In some cases, the circumstances of the integral allow you to find the exact value of the constant.

Basic integration results

REVISION NOTE

Don't worry if you can't remember all these now. All except the first are dealt with later in this book.

You need to learn the basic integration results:

For $f(x) = \ldots$	You get $\int f(x)dx = \ldots$
x^n where $n \neq -1$	$\dfrac{x^{n+1}}{n+1} + C$
$\dfrac{1}{x}$	$\ln x + C$ (or $\log_e x + C$)
$\sin x$	$-\cos x + C$
$\cos x$	$\sin x + C$
e^x	$e^x + C$

SPECIAL CASE

Notice the special case for integration of x^{-1}. All other powers of x, *including fractional powers*, follow the rule in the first line of the table.

In each case in the table above, the **arbitrary constant** that appears in the integral is shown as '$+C$'.

Integrating algebraic expressions

The table above gives the basic results. The following two rules help with expressions made from simpler functions.

Addition of two functions

REVISION NOTE

These rules just mean that integration behaves in the way you would expect. Two functions added up can be integrated separately, and multiplying by a constant doesn't need a special method.

If $\dfrac{dy}{dx} = f(x) + g(x)$ then $y = \int f(x)\, dx + \int g(x)\, dx$

Multiplication of a function by a constant

If $\dfrac{dy}{dx} = k[f(x)]$ with k constant, then $y = k\int f(x)\, dx$

These rules are equivalent to the rules for differentiation.

Indefinite integration

An integral of a function that gives a new function, including the arbitrary constant, is called an **indefinite integral**.

EXAMPLE

Find the integrals with respect to x of:

(a) $(x^3 - 3)x^2$ (b) $\sqrt[3]{x}$ (c) $\sqrt{x} + \dfrac{4}{x^3}$

(a)
$$f(x) = (x^3 - 3)x^2$$
$$= x^5 - 3x^2$$
$$\Rightarrow \int f(x)\,dx = \int x^5 - 3x^2\,dx$$
$$= \frac{x^6}{6} - x^3 + C$$

(b)
$$f(x) = \sqrt[3]{x} = x^{\frac{1}{3}}$$
$$\Rightarrow \int f(x)\,dx = \frac{x^{\frac{4}{3}}}{\left(\frac{4}{3}\right)} + C = \tfrac{3}{4}x^{\frac{4}{3}} + C$$

(c)
$$f(x) = \sqrt{x} + \frac{4}{x^3} = x^{\frac{1}{2}} + 4x^{-3}$$
$$\Rightarrow \int f(x)\,dx = \frac{x^{\frac{3}{2}}}{\left(\frac{3}{2}\right)} + 4\left(\frac{x^{-2}}{-2}\right) + C = \tfrac{2}{3}x^{\frac{3}{2}} - \frac{2}{x^2} + C$$

> **METHOD NOTE**
> We haven't yet got a method for integrating products of terms, so multiply out.

> **METHOD NOTE**
> Convert the expression into the form x^n and don't forget the arbitrary constant.

> **METHOD NOTE**
> Convert to x^n form and integrate the two terms separately.

EXAMPLE

The gradient of a graph is given at any point by the expression $3x^2 - 2$ where x is the x-coordinate.

Find the equation of the graph if it passes through $(2, 9)$.

$$\frac{dy}{dx} = 3x^2 - 2$$
$$\Rightarrow y = \int 3x^2 - 2\,dx = x^3 - 2x + C$$

So any graph of the form

$$y = x^3 - 2x + C$$

has the correct gradient at every point. Since the solution we need passes through $(2, 9)$, then

$$9 = 2^3 - 2 \times 2 + C \Rightarrow C = 5$$

So the equation of the graph is

$$y = x^3 - 2x + 5$$

> **REVISION NOTE**
> The indefinite integral, with the term '$+C$', gives a whole 'family' of graphs – all with the correct shape. See below:
>
>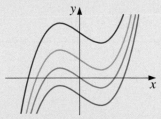
>
> Just one of these graphs passes through $(2, 9)$, and that is the one found by substitution in the final section of the example.

Area between a graph and the *x*-axis

Using integration you can find the area between a graph and the x-axis, bounded by two values of x.

To find the area, integrate the equation of the curve and substitute the two x-values into the function found. This gives a numerical value for the area.

The two values of x used are called the **limits of the integration** (one value of x is the 'lower limit' and one value is the 'upper limit'), and the method is called **definite integration**.

EXAMPLE

Find the area between the graph $y = x^2 - x + 2$, the x-axis, and the lines $x = 1$ and $x = 3$.

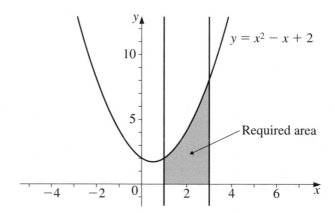

Here, the limits are the values $x = 1$ and $x = 3$. So the area is given by the definite integral

$$\text{Area} = \int_1^3 x^2 - x + 2 \, dx$$

$$= \left[\tfrac{1}{3}x^3 - \tfrac{1}{2}x^2 + 2x \right]_1^3$$

$$= \left(\tfrac{1}{3}(3)^3 - \tfrac{1}{2}(3)^2 + 2(3) \right) - \left(\tfrac{1}{3}(1)^3 - \tfrac{1}{2}(1)^2 + 2(1) \right)$$

$$= \frac{21}{2} - \frac{11}{6} = \frac{26}{3}$$

Two important points

1 If the area being calculated lies below the x-axis, the result will be negative.

It is important to check that the graph does not cross the axis in the interval between the limits of the integration. If it does, then to find the true area enclosed, you need to find the areas above and below the axes separately and add them together (see page 49).

2 If the graph has a vertical asymptote in the interval between the limits, then the area calculated is invalid.

You must check carefully that there are no 'prohibited values' for x in the interval.

Interpreting a negative result for an integral

The following example shows how to deal with areas 'below' the x-axis.

Find the area enclosed between the x-axis, the ordinates $x = 0$ and $x = 2$, and the graph $y = 3 - 3x^2$

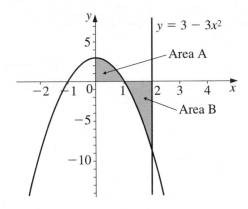

The diagram shows the graph. The area required is the sum of the two areas marked A and B.

Since area B is 'below' the x-axis, evaluating the definite integral will give a negative result, so we must integrate 'minus y'.

So, required area = area A + area B

The top limit of the integral for area A and therefore also the bottom limit of the integral for area B is the value of x where the graph crosses the x-axis.

To find this value,

$$3 - 3x^2 = 0$$
$$\implies \quad 3 = 3x^2$$
$$\implies \quad x^2 = 1$$
$$\implies \quad x = \pm 1$$

We need the crossing point between 0 and 2, so

$$\text{Area A} = \int_0^1 3 - 3x^2 \, dx$$
$$= \left[3x - x^3\right]_0^1$$
$$= (3 - 1) - (0 - 0)$$
$$= 2$$

$$\text{Area B} = -\left\{\int_1^2 3 - 3x^2 \, dx\right\}$$
$$= -\left[3x - x^3\right]_1^2$$
$$= -\{(6 - 8) - (3 - 1)\}$$
$$= -\{(-2) - (2)\}$$
$$= 4$$

So total area $= 2 + 4 = 6$ units^2

An 'ordinate' is a vertical line on a graph, joining the x-axis to the graph.

The negative result for the definite integral is *correct*. But we must interpret an *area* as a positive value.

Find where the graph crosses the x-axis by solving for $y = 0$.

The limits for area A are 0 to 1, and the limits for area B are 1 to 2.

This area is 'below' the x-axis, so we need a 'minus' sign to change the negative definite integral into a positive result for area.

The area enclosed between two curves

You can use definite integration to find the area enclosed *between* two curves.

First of all find where the graphs cross, to fix the limits of the integration. Then find each of the areas 'below' the curves, and subtract to find the enclosed area.

Find the area enclosed between the two graphs

$$y = x^2 + 1 \quad \text{and} \quad y = -x^2 + 3x + 1$$

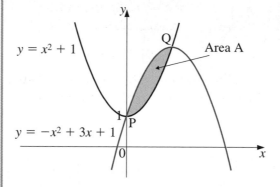

The graph shows how the two curves cross. The required area is area A.

First find the *x*-coordinates of the crossing points, marked here as P and Q.

The graphs cross when

$$x^2 + 1 = -x^2 + 3x + 1$$

$$\Rightarrow \quad 2x^2 - 3x = 0$$

$$\Rightarrow \quad x(2x - 3) = 0$$

$$\Rightarrow \quad x = 0 \quad \text{or} \quad x = \tfrac{3}{2}$$

Between the limits 0 and $\frac{3}{2}$, for the 'top' graph, $y = -x^2 + 3x + 1$,

$$\text{Area B} = \int_{0}^{\frac{3}{2}} -x^2 + 3x + 1 \, dx$$

$$= \left[-\frac{1}{3}x^3 + \frac{3}{2}x^2 + x \right]_{0}^{\frac{3}{2}}$$

$$= \left(-\frac{1}{3} \times \frac{27}{8} + \frac{3}{2} \times \frac{9}{4} + \frac{3}{2} \right) - (0)$$

$$= \frac{15}{4}$$

For the 'bottom' graph, $y = x^2 + 1$,

$$\text{Area C} = \int_{0}^{\frac{3}{2}} x^2 + 1 \, dx = \left[\frac{1}{3}x^3 + x \right]_{0}^{\frac{3}{2}}$$

$$= \left(\frac{1}{3} \times \frac{27}{8} + \frac{3}{2} \right) - (0) = \frac{21}{8}$$

So area A = area B − area C

$$= \frac{15}{4} - \frac{21}{8}$$

$$= \frac{9}{8} \text{ units}^2$$

Volume of a solid of revolution

If a graph of $y = f(x)$ is rotated about the x-axis, then a surface will be generated for which every section perpendicular to the x-axis is a circle.

This makes it particularly easy to find a formula for the volume of the solid enclosed by the surface, and limited by two values of x.

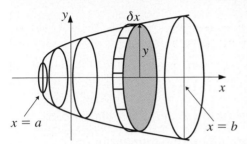

In the diagram, the blue circle has radius equal to the y-coordinate of the graph. If the 'thickness' of the slice shown is δx, then the volume of this disc is

$$\delta V = \pi y^2 . \delta x$$

So the volume of the whole solid is obtained by summing all such slices between the x-coordinates a and b, and allowing $\delta x \to 0$.

This gives a formula for the volume

$$\text{Volume} = \int_a^b \pi y^2 \, dx$$

REVISION NOTE

Learn the formula – but you are unlikely to be asked to prove it. It comes from

$$V = \lim_{\delta x \to 0} \sum_{x=a}^{x=b} \pi y^2 . \delta x$$

METHOD NOTE

If the curve is rotated about the y-axis instead, a similar formula can be proved:

$$\text{Volume} = \int_{y=c}^{y=d} \pi x^2 \, dy$$

EXAMPLE

A model of a cooling tower is the solid enclosed by rotating the graph of $y = 1 + \dfrac{x^2}{5}$ between $x = -1$ and $x = 2$ about the x-axis. Find the volume of the model.

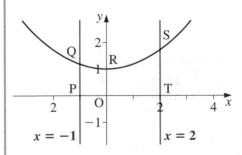

The diagram shows the graph of $y = 1 + \dfrac{x^2}{5}$, and the two lines $x = -1$ and $x = 2$. The volume is generated by rotating the shape OPQRSTO about the x-axis.

The volume is given by

$$V = \int_{-1}^{2} \pi y^2 \, dx$$

$$= \int_{-1}^{2} \pi \left(1 + \frac{x^2}{5} \right)^2 dx \quad \longleftarrow$$

$$= \pi \int_{-1}^{2} 1 + \frac{2}{5} x^2 + \frac{1}{25} x^4 \, dx$$

$$= \pi \left[x + \frac{2}{15} x^3 + \frac{1}{125} x^5 \right]_{-1}^{2}$$

$$= \pi \left\{ \left(2 + \frac{16}{15} + \frac{32}{125} \right) - \left(-1 - \frac{2}{15} - \frac{1}{125} \right) \right\} = \frac{558}{125} \pi \text{ units}^3$$

EXAM NOTE

Be careful! It's easy to forget to square the function for y.

A decimal may be easier. Make sure you answer an exam question in the appropriate style – read the question carefully.

Calculus – understanding change

1 Differentiate with respect to x:

(a) $(x^2 - 2)x^3$

(b) $\dfrac{3}{x^3} - \sqrt[3]{x}$

(c) $\sin x - \cos x$

2 For the graph

$$y = \tfrac{1}{3}x^3 - x^2 - 3x + 6$$

(a) Find the equation of the tangent with gradient -4.

(b) Find the area of the triangle enclosed between the tangent in (a) and the x and y-axes.

(c) Find the equation of the normal to the graph at the point where the tangent in (a) touches the graph.

3 Consider the graph

$$y = \frac{x^3}{3} + 3x^2 + 5x - 4$$

(a) Find the maximum and minimum points.

(b) Find where the graph crosses the y-axis, and show that when $x = -3$, the second derivative is zero.
Hence interpret this result, and sketch the graph.

4 Integrate with respect to x:

(a) $(x + 1)^2 x$

(b) $\sqrt[3]{x} + \dfrac{3}{x^2}$

(c) $\cos x + 2 \sin x$

5 Find the area enclosed between

$$y = 7 + 6x - x^2$$

the y-axis, and the positive x-axis.

6 Find the area enclosed between

$$y = x^3 - 3x^2 + x + 1,$$

the x-axis, and the ordinates $x = 0$ and $x = 2$.

7 Find the area enclosed between the two curves

$$y = x^2 - 1 \text{ and } y = -x^2 + 5x - 4.$$

8 A bowl is formed by rotating about the x-axis part of the curve

$$y = 3 + \sqrt{x}$$

between $x = 0$ and $x = 4$.
Find the volume of the bowl, leaving your answer as a multiple of π.

9 Find the volume enclosed by rotating the section of the graph

$$y = x^2$$

between the origin and $(2, 4)$, about the y-axis.

6 Differentiation and integration

The chain rule

All of the following expressions

$$(x^3 + 2)^7, \quad \sqrt{x^2 - 1}, \quad \sin(x^2), \quad e^{x^2}$$

are example of **composite functions**. To evaluate them for a particular value of x, you would need to work out the outcome of one function, and then use a second function to arrive at the final answer.

For example, for the expression $(x^3 + 2)^7$ at $x = 2$, if $y = (x^3 + 2)^7$ then the composite function can be broken down into two steps:

1 $u = x^3 + 2$ and **2** $y = u^7$.

So,

$$\begin{aligned} x = 2 \Rightarrow u &= 2^3 + 2 \\ &= (x^3 + 2) = 10 \\ \Rightarrow y &= u^7 = 10^7. \end{aligned}$$

The chain rule

For differentiating a composite function $y = f(x)$

$$\frac{dy}{dx} = \frac{dy}{du} \times \frac{du}{dx}$$

REVISION NOTE

$(x^3 + 2)^7$ could be done by multiplying the bracket out. But it would be messy – the chain rule is the best method! The others, of course cannot be multiplied out.

NOTE

The exponential function is dealt with on page 63.

METHOD NOTE

The choice of 'u' as the extra variable is not important. Any variable will do.

EXAMPLE

Differentiate $y = (x^3 + 2)^7$ with respect to x.

$y = (x^3 + 2)^7$ can be split into $u = x^3 + 2$ and $y = u^7$

So $\dfrac{dy}{du} = 7u^6$

$\qquad = 7(x^3 + 2)^6$

and $\dfrac{du}{dx} = 3x^2$

So, using the chain rule,

$$\begin{aligned} \frac{dy}{dx} &= \frac{dy}{du} \cdot \frac{du}{dx} \\ &= [7(x^3 + 2)^6](3x^2) \qquad \text{(A)} \\ &= 21x^2(x^3 + 2)^6 \end{aligned}$$

METHOD NOTE

Differentiate using 'u' and then put the expression for 'x' back in.

Now multiply the two separate parts for $\dfrac{dy}{du}$ and $\dfrac{du}{dx}$. Finally, 'tidy up' the product to give the answer.

EXAM NOTE

You *could* have found this answer by multiplying out, differentiating, and factorising again, but it's really not recommended.

With practice you can often write down directly the step labelled (A) in the example above. In an exam it is better to show the intermediate steps introducing the function u, so that if you make a mistake in the calculation you will still gain method marks.

Using the chain rule

EXAMPLE

Differentiate $y = \sqrt{x^2 - 1}$ with respect to x.

$y = \sqrt{x^2 - 1}$ can be split into $u = x^2 - 1$ and $y = \sqrt{u}$

So $\dfrac{dy}{du} = \dfrac{1}{2}u^{-\frac{1}{2}} = \dfrac{1}{2\sqrt{u}}$

$\qquad = \dfrac{1}{2\sqrt{x^2 - 1}}$

and $\dfrac{du}{dx} = 2x$

So, using the chain rule,

$$\frac{dy}{dx} = \frac{dy}{du} \cdot \frac{du}{dx}$$

$$= \left[\frac{1}{2\sqrt{x^2 - 1}}\right] \cdot (2x)$$

$$= \frac{x}{\sqrt{x^2 - 1}}$$

METHOD NOTE

Separate the composite function into two that are easy to differentiate separately.

METHOD NOTE

After differentiating $y = f(u)$, replace 'u' by the expression for x.

Combine the two derivatives, and tidy up the new expression for the whole derivative.

Calculating rates of change for linked variables

The chain rule also provides a method for calculating rates of change for linked variables.

EXAMPLE

A slag heap in a quarry forms a conical pile, with height equal to the radius of the cone. When the conical heap is 10 metres high, new slag is being added to it by a conveyor belt at a rate of 1 cubic metre per second.

How fast is the height of the heap growing?

The volume of a cone is $V = \frac{1}{3}\pi r^2 h$, and here $r = h$. The slag being added by the conveyor belt means $\dfrac{dV}{dt} = 1$ and you have been asked to find $\dfrac{dh}{dt}$, when $h = 10$.

Here, $r = h \implies V = \frac{1}{3}\pi h^3$

$$\implies \frac{dV}{dh} = \pi h^2$$

By the chain rule, $\dfrac{dV}{dt} = \dfrac{dV}{dh} \times \dfrac{dh}{dt} = \pi h^2 \times \dfrac{dh}{dt}$

so $\dfrac{dh}{dt} = \dfrac{1}{\pi h^2} \cdot \dfrac{dV}{dt}$

and $h = 10$, $\dfrac{dV}{dt} = 1 \implies \dfrac{dh}{dt} = \dfrac{1}{\pi(100)} \times 1 = 0.003\,18$ (3 s.f)

So the heap is growing at 3 mm per second at this instant.

EXAM NOTE

Work out first from the question:
What do I know? What do I have to find? What formula connects them?
You usually need to use *all* the information you have been given.

METHOD NOTE

This example has 3 variables: V, r and h. So we have to use the link to get rid of r, as we need to work with V and h.

This is the line that makes it all work.

EXAM NOTE

Don't lose marks for units. This question is in metres and seconds, so the answer is 0.003 metres per second which is 3 mm per second.

Integration by inspection using the chain rule

The standard form of the answer you get when using the chain rule can sometimes help you to spot a short cut to an integral.

An alternative way of writing the chain rule helps to understand the trick:

If $y = f(g(x))$ then $y = f(u)$ and $u = g(x)$

So $\dfrac{dy}{dx} = \dfrac{dy}{du} \times \dfrac{du}{dx}$

$= f'(u) \times g'(x) = g'(x) \cdot f'(g(x))$

When you are integrating an expression, if you can spot that it consists of the product of a function and the derivative of that function you can often 'guess' the answer.

EXAMPLE

Evaluate the integral with respect to x of the expression $2x(x^2 - 1)^2$.

Here, we 'spot' that the derivative of $x^2 - 1$ is $2x$. So the expression to be integrated seems to be in the right form for the derivative of $(x^2 - 1)^3$.

Differentiate this, to check the outcome.

If $y = (x^2 - 1)^3$, it can be split into

$y = u^3$ and $u = x^2 - 1$

so that $\dfrac{dy}{du} = 3u^2 = 3(x^2 - 1)^2$

and $\dfrac{du}{dx} = 2x$

By the chain rule, $\dfrac{dy}{dx} = \dfrac{dy}{du} \times \dfrac{du}{dx}$

so $\dfrac{d}{dx}[(x^2 - 1)^3] = 3(x^2 - 1)^2 \cdot (2x)$

$= 6x(x^2 - 1)^2$

Reversing this differentiation, we can write down the answer

$\int 2x(x^2 - 1)^2 \, dx = \tfrac{1}{3}\int 6x(x^2 - 1)^2 \, dx$

$= \tfrac{1}{3}(x^2 - 1)^3 + C$

This result could be found by multiplying out the bracket in the question and then integrating and factorising, but this direct approach is easier.

There are three steps to integration by inspection using the chain rule.

1 Spot that the derivative of one term looks like the other.
2 Differentiate an expression based on the first term.
3 Adjust the outcome of the differentiation to match the question.

Check out these steps in the example above.

REVISION NOTE

This expression looks complicated – you don't need to remember it!

Think of this as 'derivative of the bracket' times 'derivative of what's inside the bracket'.

REVISION NOTE

This method is 'on the fringe' of most AS-Level syllabuses. Check past papers to see if it would have helped.

METHOD NOTE

The initial expression looks like 'something squared'. So try differentiating 'something cubed'.

NOTE

Follow the example through – it's good differentiation practice of the chain rule.

This has the correct form, though it is $6 \times \ldots$ instead of $2 \times \ldots$ So we divide by 3 on both sides to get the answer.

NOTE

xe^x is dealt with on page 66.

METHOD NOTE

In some products, you may need to use the chain rule to differentiate the separate terms. See the examples on page 57.

REVISION NOTE

This proof should help you to understand how the product rule works. But you will not be asked to write out a proof in AS-Level, so don't try to learn it by heart.

METHOD NOTE

These terms are represented by the 4 parts of the diagram. Next subtract the original $y = u.v$ from both sides.

This step relies on our definition of $\dfrac{dy}{dx}$ as the limit of the increment in y, divided by the increment in x – see Unit 4.

The product rule

The expressions

$$(x^3 - 1)(x^2 + 2x - 5), \ (x^2 - 1)\sqrt{2x + 1}, \ x^2 \sin x, \ xe^x$$

are all examples of **products**. Each is constructed from two separate, easy to differentiate functions, multiplied together.

All of these expressions, and many similar ones can be differentiated using the product rule.

The product rule

If $y = u(x).v(x)$ where u and v are functions of x,

then $\dfrac{dy}{dx} = u\dfrac{dv}{dx} + v\dfrac{du}{dx}$

There is a simple 'graphical' demonstration of this result, using the area of a rectangle to represent the product $u.v$.

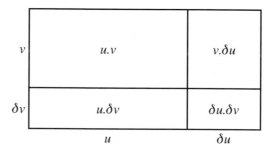

If $y = u.v$ where u and v are functions of x, then the area of the outer rectangle in the diagram can represent the value of the function when u and v are slightly increased to $u + \delta u$ and $v + \delta v$ by a small increase in x, say δx.

So we can write the area in the form:

$$y + \delta y = (u + \delta u) \times (v + \delta v)$$

$$\Rightarrow \quad uv + \delta y = uv + u.\delta v + v.\delta u + \delta u.\delta v$$

$$\Rightarrow \quad \delta y = u.\delta v + v.\delta u + \delta u.\delta v$$

Dividing throughout by δx, and then considering what happens as δx tends to zero,

$$\frac{\delta y}{\delta x} = \frac{u\delta v}{\delta x} + \frac{v\delta u}{\delta x} + \frac{\delta u\delta v}{\delta x}$$

and as δx tends to zero, the last term tends to zero while each of the other terms gives a derivative, since, for example,

$$\lim_{\delta x \to 0} \left(u\frac{\delta v}{\delta x} \right) = u \times \lim_{\delta x \to 0} \frac{\delta v}{\delta x} = u\frac{dv}{dx}$$

So we get

$$\frac{dy}{dx} = u\frac{dv}{dx} + v\frac{du}{dx}$$

which is the product rule.

Using the product rule

EXAMPLE

Differentiate with respect to x:

(a) $(x^3 - 1)(x^2 + 2x - 5)$

(b) $(x^2 - 1)\sqrt{2x + 1}$

(a) $y = (x^3 - 1)(x^2 + 2x - 5)$

Put $u = x^3 - 1$ and $v = x^2 + 2x - 5$

$\Rightarrow \dfrac{du}{dx} = 3x^2$ and $\dfrac{dv}{dx} = 2x + 2$

So $\dfrac{dy}{dx} = u\dfrac{dv}{dx} + v\dfrac{du}{dx}$

$\qquad = (x^3 - 1)(2x + 2) + (x^2 + 2x - 5)(3x^2)$

$\qquad = (2x^4 + 2x^3 - 2x - 2) + (3x^4 + 6x^3 - 15x^2)$

$\qquad = 5x^4 + 8x^3 - 15x^2 - 2x - 2$

> **METHOD NOTE**
>
> This is a product, so separate the terms. Each of u and v is easy to differentiate separately.

> There's some nasty algebra here. This one doesn't factorise, but many will.

(b) $y = (x^2 - 1)\sqrt{2x + 1}$

Put $u = x^2 - 1$ and $v = \sqrt{2x + 1}$

$\Rightarrow \dfrac{du}{dx} = 2x$ and $\dfrac{dv}{dx} = \dfrac{2}{2\sqrt{2x + 1}}$

$\qquad\qquad\qquad\qquad = \dfrac{1}{\sqrt{2x + 1}}$

So $\dfrac{dy}{dx} = (x^2 - 1)\left(\dfrac{1}{\sqrt{2x + 1}}\right) + (\sqrt{2x + 1})(2x)$

$\qquad = \dfrac{(x^2 - 1) + (2x + 1)(2x)}{\sqrt{2x + 1}}$

$\qquad = \dfrac{(x^2 - 1) + (4x^2 + 2x)}{\sqrt{2x + 1}}$

$\qquad = \dfrac{5x^2 + 2x - 1}{\sqrt{2x + 1}}$

> **METHOD NOTE**
>
> The differentiation of v uses the chain rule – check it.

> **EXAM NOTE**
>
> Some calculus questions in the AS exams seem to turn into tests of your algebra – like this one. Check through the working here to make sure you understand it.

EXAMPLE

Find the gradient of $y = \frac{1}{4}x^2 \sin x$ when $x = \dfrac{\pi}{2}$

If $u = \frac{1}{4}x^2$ and $v = \sin x$

then $\dfrac{du}{dx} = \frac{1}{2}x$ and $\dfrac{dv}{dx} = \cos x$

so $\dfrac{dy}{dx} = \frac{1}{4}x^2 \cos x + \frac{1}{2}x \sin x$

$x = \dfrac{\pi}{2} \Rightarrow \dfrac{dy}{dx} = \dfrac{1}{4}\dfrac{\pi^2}{4}(0) + \dfrac{1}{2}\dfrac{\pi}{2}(1) = \dfrac{\pi}{4}$

> **REVISION NOTE**
>
> The graph and tangent look like this:
>
>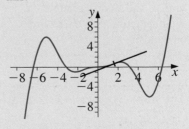

> So the gradient is approximately 0.785 (3 s.f)

NOTE

$\dfrac{e^x}{x^3}$ is dealt with on page 66.

The quotient rule

You may need to differentiate expressions like these:

$$\frac{x+1}{x-3}, \quad \frac{x-1}{\sqrt{x^2+3}}, \quad \frac{\sin x}{x}, \quad \frac{e^x}{x^3}$$

All of these are **quotients** – one expression divided by another. They can be differentiated using the product and chain rules. For example, rewriting $\dfrac{x+1}{x-3}$ as a product in the form

$$(x+1)(x-3)^{-1}$$

means you can use the product rule. The second bracket will need to be differentiated using the chain rule.

But it makes sense to use the same idea on a general case to get a new rule, called the **quotient rule**.

REVISION NOTE

Try to understand this proof if it helps. You will not be asked to prove it in an exam, so just learn the result.

If $y = \dfrac{u}{v}$ where u and v are functions of x, then $y = u.\left(\dfrac{1}{v}\right)$

Treating this as a product, differentiating u gives $\dfrac{du}{dx}$, and differentiating $\dfrac{1}{v}$ using the chain rule on v^{-1} gives

$-\dfrac{1}{v^2} \times \dfrac{dv}{dx}$. So the product rule gives

METHOD NOTE

The chain rule here gives

$$\frac{d}{dx}(v^{-1}) = \frac{d}{dv}(v^{-1}) \times \frac{dv}{dx}$$

$$= -\frac{1}{v^2} \times \frac{dv}{dx}$$

$$\frac{dy}{dx} = u\left(-\frac{1}{v^2} \cdot \frac{dv}{dx}\right) + \left(\frac{1}{v}\right) \cdot \frac{du}{dx}$$

$$= \frac{v \cdot \dfrac{du}{dx} - u \cdot \dfrac{dv}{dx}}{v^2}$$

The quotient rule

If $y = \dfrac{u}{v}$ where u and v are functions of x,

then $\dfrac{dy}{dx} = \dfrac{v \cdot \dfrac{du}{dx} - u \cdot \dfrac{dv}{dx}}{v^2}$

REVISION NOTE

For some quotient rule questions you will need to use the chain rule as well to differentiate the separate terms. See the example on page 59.

Although the formula looks more complicated than the product rule, most people find it easier to use this direct method.

Look at the formula carefully – it is not symmetric. The first term in the numerator contains the denominator of the original function.

It is usually best to show the intermediate steps of finding $\dfrac{du}{dx}$ and $\dfrac{dv}{dx}$ before quoting the quotient rule formula, so that you can still gain method marks if you make a mistake in your working.

Using the quotient rule

Use the quotient rule to differentiate with respect to x:

(a) $\dfrac{x^2 + 1}{4x - 3}$ (b) $\dfrac{5x + 3}{\sqrt{x^2 - 6}}$

These two examples show how to apply the quotient rule. But exam questions can become a test of your algebra.

(a) If $y = \dfrac{x^2 + 1}{4x - 3}$ then the expression can be written

as $y = \dfrac{u}{v}$ where $u = x^2 + 1$ and $v = 4x - 3$

Here there are two simple expressions to differentiate.

$$\Rightarrow \qquad \frac{du}{dx} = 2x \quad \text{and} \quad \frac{dv}{dx} = 4$$

The quotient rule gives

$$\frac{dy}{dx} = \frac{v\dfrac{du}{dx} - u\dfrac{dv}{dx}}{v^2}$$

$$= \frac{(4x - 3)(2x) - (x^2 + 1)(4)}{(4x - 3)^2}$$

$$= \frac{2(x^2 - 3x - 2)}{(4x - 3)^2}$$

(b) $y = \dfrac{5x + 3}{\sqrt{x^2 - 6}} \Rightarrow y = \dfrac{u}{v}$

This example does not separate into such simple expressions. Differentiating v needs the chain rule.

where $u = 5x + 3$, $v = \sqrt{x^2 - 6}$

$$\Rightarrow \quad \frac{du}{dx} = 5 \quad \text{and} \quad \frac{dv}{dx} = \tfrac{1}{2}(x^2 - 6)^{-\frac{1}{2}} \cdot (2x)$$

$$= \frac{x}{(x^2 - 6)^{\frac{1}{2}}}$$

Check the use of the chain rule here, and some tidying up of the derivative.

$$\frac{dy}{dx} = \frac{v\dfrac{du}{dx} - u\dfrac{dv}{dx}}{v^2}$$

And here is the algebra test. Notice how the 'square root' term is dealt with.

$$\Rightarrow \quad \frac{dy}{dx} = \frac{(x^2 - 6)^{\frac{1}{2}}(5) - (5x + 3)\left(\dfrac{x}{(x^2 - 6)^{\frac{1}{2}}}\right)}{(\sqrt{x^2 - 6})^2}$$

$$= \frac{(x^2 - 6)(5) - (5x + 3)x}{(x^2 - 6)^{\frac{3}{2}}}$$

$$= \frac{5x^2 - 30 - 5x^2 - 3x}{(x^2 - 6)^{\frac{3}{2}}}$$

Then factorise as much as possible to give the simplest answer.

$$= \frac{-3(10 + x)}{(x^2 - 6)^{\frac{3}{2}}}$$

Using the quotient rule (cont.)

NOTE

What happens to this graph when $x = 0$?

Both $\sin x$ and x are zero for $x = 0$, so the value of y is undefined. This is a special case: if you take values ever closer to zero, $\sin x$ and x get closer and closer in value – you could try this on your calculator!

So the expression $\dfrac{\sin x}{x} \to 1$.

This means that here the value of y can be taken to be 5.

Take a look at the graph below the example. You don't need this result for AS-Level, but it is interesting.

EXAMPLE

Find the gradient of the curve

$$y = \frac{5 \sin x}{x}$$

at the first two points where the graph crosses the x-axis where the x-coordinate is positive.

The graph crosses the axis when

$$\frac{5 \sin x}{x} = 0 \implies \sin x = 0$$
$$\implies \pi, 2\pi, 3\pi, \ldots$$

So if $y = \dfrac{u}{v}$ where $u = 5 \sin x$ and $v = x$

then $\dfrac{du}{dx} = 5 \cos x$ and $\dfrac{dv}{dx} = 1$

The quotient rule gives

$$\frac{dy}{dx} = \frac{v\dfrac{du}{dx} - u\dfrac{dv}{dx}}{v^2}$$

$$= \frac{x(5 \cos x) - (5 \sin x).1}{x^2}$$

$$= \frac{5(x \cos x - \sin x)}{x^2}$$

When $x = \pi$, then $\dfrac{dy}{dx} = \dfrac{5(\pi \cos \pi - \sin \pi)}{\pi^2}$

$$= \frac{5(\pi(-1) - 0)}{\pi^2} = -\frac{5}{\pi}$$

When $x = 2\pi$, then $\dfrac{dy}{dx} = \dfrac{5(2\pi \cos 2\pi - \sin 2\pi)}{(2\pi)^2}$

$$= \frac{5(2\pi(1) - (0))}{4\pi^2} = \frac{5}{2\pi}$$

So the gradient at the first intercept with the x-axis is -1.59 (3 s.f.) and at the second is $+0.796$ (3 s.f.)

In fact, the graph looks like:

The diagram shows the graph and the two tangents at the first two crossing points on the positive x-axis.

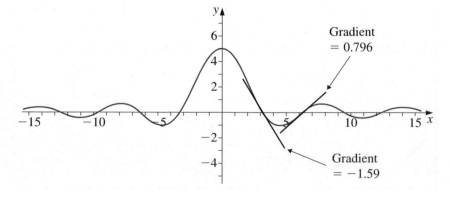

Differentiation and integration

1 Use the chain rule, product rule and quotient rule as appropriate to differentiate with respect to x:

(a) $(x^3 + 2)^6$

(b) $\dfrac{x - 1}{x + 2}$

(c) $(x + 1)(x^3 - 1)$

(d) $\sqrt{x^3 + 1}$

(e) $x \cos x$

(f) $\dfrac{\cos x}{\sqrt{x}}$

(g) $x^2 e^{-x}$

(h) $(x - 2)\sqrt{x + 1}$

2 A spherical weather balloon is inflated by pumping in gas at a rate of $2 \, \text{m}^3$ per second. Find the rate of increase of the radius of the balloon, measured in centimetres per second, when the balloon has a radius of 3 metres.

3 Integrate with respect to x, by inspection using the chain rule:

(a) $x^2(x^3 + 1)^5$

(b) $\dfrac{x^2}{\sqrt{x^3 - 1}}$

4 The formula $s = ut + \frac{1}{2}gt^2$ gives the vertical displacement s of a cliff-diver from the end of a diving board, after t seconds. In the formula, u and g are constants equal to $+2$ and -10 respectively.

(a) Show that the diver enters the water 16 metres below the end of the board after 2 seconds.

(b) The velocity of the diver is the derivative of s with respect to t. Find $\dfrac{ds}{dt}$ and find the speed of the diver when she enters the water.

5 Find $\int \sin x \cos^2 x \, dx$

6 By writing $\sec x$ as $\dfrac{1}{\cos x}$ and using the chain rule, show that
$$\frac{d}{dx}(\sec x) = \sec x \tan x$$

7 Differentiate with respect to x:

(a) $e^x \cos x$

(b) $\dfrac{x^2}{x - 1}$

(c) $(x^2 - 3)^4$

(d) $\sin x - \cos x$

8 Integrate with respect to x, by inspection using the chain rule:

(a) $x(x^2 - 3)^3$

(b) $\dfrac{\sin x}{\cos^2 x}$

(c) Use your result from (b) to find
$$\int_0^{\frac{\pi}{4}} \frac{\tan x}{\cos x} \, dx$$

7 Logs and exponentials

Definition of a log

REVISION NOTE

Here 'b' is the **base** of the logarithm.

So here, log a is the power that the base b is raised to in order to equal a.

If $a = b^x$ then $\log_b a = x$

This simple rule shows how **logarithms** and **indices** (powers) are linked.

For example,

$$1000 = 10^3 \implies \log_{10} 1000 = 3$$

and $\quad \frac{1}{8} = 2^{-3} \implies \log_2 \frac{1}{8} = -3$

Notice also, for example, that

$$3 = \sqrt{9} = 9^{\frac{1}{2}}$$

$$\implies \quad \log_9 3 = \tfrac{1}{2} = 0.5$$

METHOD NOTE

Here the bases are 10 and 2 respectively. The usual ones are 10 or 'e'. (See page 63.)

Here, 3 is the square root of 9, or 9 to the power one-half.

These two examples show how negative and fractional logs arise.

Indices and the laws of logs

NOTE

These three rules work *whichever* base you choose – just as long as you are consistent.

The three fundamental laws of logarithms are:

$$\log a + \log b = \log (ab)$$

$$\log a - \log b = \log \left(\frac{a}{b}\right)$$

$$k \log a = \log (a^k)$$

REVISION NOTE

You really only need to learn the first rule.

The second follows directly if you use

$+\log \dfrac{1}{b} = -\log b$ and the third follows directly from the first if $b = a$.

Solving problems with logs

EXAMPLE

Simplify $2 \log x - 4 \log y$ by writing it as a logarithm of a single expression.

$$2 \log x - 4 \log y = \log (x^2) - \log (y^4)$$

$$= \log \left(\frac{x^2}{y^4}\right)$$

EXAMPLE

John starts a rumour spreading through his school. The number of people who have heard the rumour can be approximated by $n = 3^t$ after t hours. After how long will 100 people have heard the rumour?

We need to solve

$$3^t = 100$$

$$\implies \quad \log (3^t) = \log (100)$$

$$\implies \quad t \log (3) = \log (100)$$

$$\implies \qquad t = \frac{\log (100)}{\log (3)} = \frac{2}{0.4771} = 4.19 \text{ (3 s.f.)}$$

So 100 people will have heard the rumour after about 4.2 hours.

METHOD NOTE

Here we have used logs to base 10, which are on your calculator. But we could have used a different base.

If we had used base 'e' (also on your calculator) the sum would have been

$$\frac{\log (100)}{\log (3)} = \frac{4.605\ldots}{1.098\ldots} = 4.191\ldots$$

as before.

Exponential functions

On page 62, $\log_b a = x$ was defined in terms of the expression $a = b^x$.

> Functions of the form $y = k^x$, where k is a positive constant, are called **exponential functions**.

The diagram shows the graphs of a set of exponential functions, for different values of k.

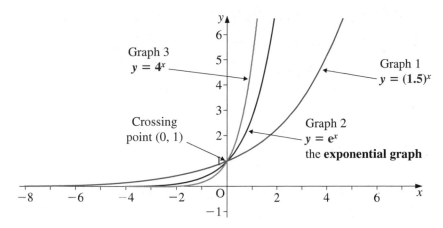

Graph 3
$y = 4^x$

Graph 1
$y = (1.5)^x$

Crossing point (0, 1)

Graph 2
$y = e^x$
the **exponential graph**

REVISION NOTE

This graph is not defined if k is a negative number. There are some values of x where k^x does not exist for negative k, e.g. for $x = 0.5$ (square root of negative number)

NOTE

$x \rightarrow +\infty \Rightarrow y \rightarrow +\infty$
$x \rightarrow -\infty \Rightarrow y \rightarrow 0$

NOTE

Because for any positive value of k, then $k^0 = 1$.

Notice that they all pass through (0, 1) and they are all similar in shape. The value of k determines the 'steepness'.

By varying k, any positive gradient at (0, 1) can be obtained from $f(x) = k^x$ where k is a positive constant.

For just one of those values of k, a special case occurs where the gradient of the graph as it passes through the point (0, 1) equals 1.

> For this special value of k, the gradient of the graph as it passes through (0, 1) is 1.
>
> $\quad k = e = 2.718\,281\,828\,46...$
>
> and the graph is $y = e^x$.
>
> This is often called **the exponential function**.

REVISION NOTE

It looks as though this value of 'e' is a recurring decimal, but it is just coincidental that there is a pattern in these first few terms. Like all irrational numbers, it *never* repeats.

This value e is an important constant. It is an irrational number, but can be obtained to many decimal places from your calculator. It has many important applications in calculus, and is the 'base' that mathematicians use for **natural logarithms** – that is 'logs to base e', usually written as ln (x).

Some people prefer $\log_e (x)$ instead of ln (x).

The exponential function and e

The constant e turns up in many branches of mathematics, but the simplest formula to generate its numerical value is the **exponential series**:

REVISION NOTE

Make sure you know how to get e^x and ln (x) from your calculator.

$$e^x = 1 + x + \frac{x^2}{2} + \frac{x^3}{6} + ... + \frac{x^n}{n!} + ...$$

Putting $x = 1$ into this series gives the value $e^1 = e = 2.718\,281\,8...$ but many terms are needed to get this value.

Negative exponentials

The graphs of $y = k^x$ are shown on page 63. From those graphs it is easy to work out the graph of $y = k^{-x} = \dfrac{1}{k^x}$, where k is a positive constant.

The diagram shows some **negative exponential graphs**.

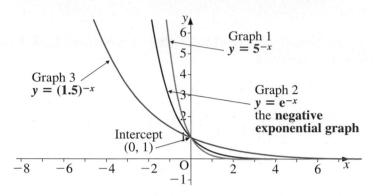

The graphs all pass through $(0, 1)$ and are mirror images in the y-axis of the graphs of $y = k^x$ on page 63.

For these graphs of $y = k^{-x}$

$$x \to +\infty \Rightarrow y \to 0$$
$$x \to -\infty \Rightarrow y \to +\infty$$

which is opposite to the graphs of $y = k^x$.

Also, for all values of x greater than zero, the value of y is less than 1, while for all values of x less than zero, the value of y is greater than 1.

Exponentials and logs

The functions $y = k^x$ and $y = \log_k (x)$ are inverse functions. The graphs of the two functions are mirror images of each other in the line $y = x$.

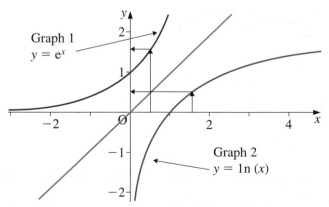

In the diagram, for the graph $y = e^x$,

$$x = \tfrac{1}{2} \Rightarrow y \approx 1.65$$

and then for $y = \ln (x)$,

$$x = 1.65 \Rightarrow y \approx 0.5$$

as expected for inverse functions.

This example uses exponentials and logs to base e, but a similar relationship follows for other exponents and logarithm bases.

Exponentials, logs and calculus

The choice of the value of e so that the gradient of the graph of $y = e^x$ at the point $(0, 1)$ is 1 means that this exponential graph has another very important property:

> If $y = e^x$ then $\dfrac{dy}{dx} = e^x$.
>
> So $\int e^x \, dx = e^x + C$, where C is the constant of integration.

REVISION NOTE

This result means that the gradient at any point on $y = e^x$ is equal to the y-coordinate.

It also follows directly that:

> If $y = \ln x$ then $\dfrac{dy}{dx} = \dfrac{1}{x}$
>
> and $\int \dfrac{1}{x} \, dx = \ln |x| + C$

EXAM NOTE

This result must be used with great care. The graph of $y = \dfrac{1}{x}$ has a prohibited value at $x = 0$, where the graph goes off to infinity. You must be careful that the interval between the limits of any integral does not include $x = 0$.

These results were included on pages 38 and 46.

All the differentiation and integration methods described in Units 5 and 6 can be used with exponentials and logs.

Differentiating and integrating exponentials

EXAMPLE

Find the derivative with respect to x of e^{x^2}.

The expression $y = e^{x^2}$ is a **composite** function, made up of the functions

$$y = e^u \quad \text{and} \quad u = x^2$$

So we can find the derivative by using the chain rule

Here, $y = e^u$ and $u = x^2$

$$\Rightarrow \quad \frac{dy}{du} = e^u \quad \Rightarrow \quad \frac{du}{dx} = 2x$$

$$= e^{x^2}$$

So, by the chain rule,

$$\frac{dy}{dx} = \frac{dy}{du} \times \frac{du}{dx}$$

$$= e^{x^2} \times 2x$$

$$= 2xe^{x^2} \longleftarrow$$

REVISION NOTE

Look at page 53 to remind yourself of how the chain rule works.

EXAM NOTE

It's clearer to write the final answer in this format, reversing the order of the terms. There is then no possible confusion about what is the power.

Graph of $y = e^{x^2}$

If the question had required the *gradient* of the graph, you could use this result, and substitute a particular value of x. You could then use the methods from Unit 2, **Coordinate geometry**, to find the equation of a tangent or normal, for example.

Differentiating and integrating exponentials (cont.)

EXAMPLE

Find the area between the graph of $y = \dfrac{1}{1 + x}$, the x-axis, the y-axis, and the line $x = 1$.

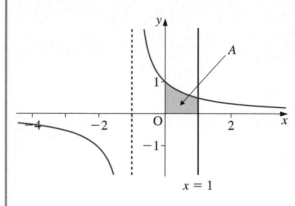

It is important to check that the graph does not cross the axis in the interval we will use, and that there are no prohibited values of x. The graph is shown here, with the required area marked A.

Since $\dfrac{d}{dx} (\ln (1 + x)) = \dfrac{1}{1 + x}$

then $A = \displaystyle\int_0^1 \dfrac{1}{1 + x} \, dx = \Big[\ln(1 + x)\Big]_0^1$

$= \ln (2) - \ln (1) = \ln (2)$

REVISION NOTE

This is really a chain rule differentiation, with $y = \ln (u)$, where $u = 1 + x$.

IMPORTANT

Remember $\ln (1) = 0$.

Look back at the graphs of logs – they all go through $(1, 0)$.

EXAMPLE

Differentiate with respect to x:

(a) $y = xe^x$ (b) $y = \dfrac{e^x}{x^3}$

(a) $y = xe^x = u.v$

where $u = x$ and $v = e^x$

$\Rightarrow \quad \dfrac{du}{dx} = 1$ and $\dfrac{dv}{dx} = e^x$

So $\quad \dfrac{dy}{dx} = u\dfrac{dv}{dx} + v\dfrac{du}{dx}$

$= xe^x + e^x.1$

$= e^x(x + 1)$

(b) $y = \dfrac{e^x}{x^3} = \dfrac{u}{v}$

where $u = e^x$ and $v = x^3$

$\Rightarrow \quad \dfrac{du}{dx} = e^x \Rightarrow \dfrac{dv}{dx} = 3x^2$

So $\quad \dfrac{dy}{dx} = \dfrac{v\dfrac{du}{dx} - u\dfrac{dv}{dx}}{v^2}$

$= \dfrac{x^3e^x - e^x(3x^2)}{(x^3)^2}$

$= \dfrac{e^x (x - 3)}{x^4}$

METHOD NOTE

Use the product rule with the product split into two terms, each of which can be differentiated simply.

Use the quotient rule.

REVISION NOTE

Check through the 'tidying-up' algebraic manipulation here.

Logs and exponentials

1 Solve for x:

    (a) $5^x = 1000$ (b) $\log_x 4 = 2$ (c) $\ln(2^x) = 10$

2 Find the area between

$$y = \frac{2}{3 + x}$$

    the x-axis, the y-axis and the line $x = 5$.

3 Differentiate with respect to x:

    (a) $x^2 e^{-3x}$ (b) $\left(\dfrac{1}{x^4}\right) e^{2x}$

4 The curve

$$y = kx^n \quad (1)$$

    passes through A(1, 5.21) and B(6, 9.33), where the y-coordinates are approximate, but given to two decimal places.

    (a) Show that by taking logarithms in equation (1) you can obtain the equation

$$\log_{10} y = n \log_{10} x + \log_{10} k \quad (2)$$

    (b) Use point A and equation (2) to estimate an approximate value for k.

    (c) Use point B and equation (2) to estimate an approximate value for n, and hence write down the equation of the curve.

5 A curve of the form

$$y = Ax^n \quad (1)$$

    is believed to represent the relationship between two variables, where values are known only approximately.

    (a) Explain why changing the equation (1) to the form

$$\log_{10} y = n \log_{10} x + \log_{10} A$$

    will transform the data for x and y into a form that should produce a straight line graph when $\log_{10} y$ is plotted against $\log_{10} x$.

    (b) Complete the table:

x	1	2	3	4	5	6
y	2.21	6.39	8.76	11.67	18.41	20.44
$\log_{10} x$						
$\log_{10} y$						

    (c) Plot the values of $\log_{10} y$ against the values of $\log_{10} x$ and draw a line of best fit by eye. Hence find suitable values of A and n in equation (1).

8 Numerical methods

An important branch of mathematics deals with numerical methods – finding approximate solutions to an appropriate degree of accuracy by arithmetical calculation. Good numerical methods are not simply 'number crunching' – they involve carefully planned strategies to solve problems that are not easy to solve using analytical approaches.

Solving equations by numerical methods

All numerical methods start with finding an approximate **initial estimate** for the root, or finding an **interval** in which the solution lies.

For this graph, the smallest root of the equation lies in the interval $-2 < x < -1$.

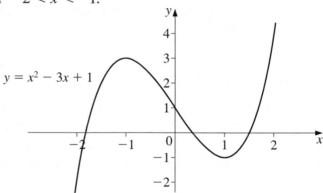

To find the roots you can use a **change of sign** method. This relies on searching for a change of sign in $y = f(x)$.

Change of sign methods

> For a continuous graph, if you can find an interval $a < x < b$ where $f(a)$ and $f(b)$ have opposite signs, then there must be a value between a and b where the graph of $y = f(x)$ crosses the x-axis.

NOTE

This result is true only if the graph is continuous between a and b.

If the graph goes off to infinity, for example, it may *not* cross the axis in the interval.

Decimal search method

EXAMPLE

Find the smallest solution of $f(x) = x^3 - 3x + 1$

$f(-2) < 0$ and $f(-1) > 0$, so there is a root in $-2 < x < -1$

Now we search in steps of 0.1, starting with $x = -2$, then $x = -1.9$, and so on, until we find a change of sign. It is easy to complete a table:

METHOD NOTE

You don't really need the actual values of $f(x)$ – just the sign. See the next table.

x	−2.0	−1.9	−1.8	−1.7	−1.6	−1.5	−1.4	−1.3	−1.2	−1.1	−1.0
$f(x)$	−1.00	−0.16	0.57	1.19	1.70	2.13	2.46	2.70	2.87	2.97	3.00

So now we can see that $f(-1.9) < 0$ and $f(-1.8) > 0$, so there is a root in $-1.9 < x < -1.8$.

This process can be repeated, now checking in steps of 0.01 between -1.90 and -1.80 until we find the root to a suitable degree of accuracy.

So, creating another table:

x	-1.90	-1.89	-1.88	-1.87	-1.86	-1.85	-1.84	-1.83	-1.82	-1.81	-1.80
$f(x)$	<0	<0	<0	>0	>0	>0	>0	>0	>0	>0	>0

This time, the sign change comes in the interval

$$-1.88 < x < -1.87$$

Another **iteration** on this interval, using an increment of 0.001, tells us that the root lies in $-1.880 < x < -1.879$.

When we have sufficient accuracy, we choose the **midpoint** of the interval as our **best estimate** of the root.

So we estimate the smallest root of the equation to be

$$x = -1.8795 \pm 0.0005$$

You could find the remaining roots of the equation by starting with a different interval.

> **METHOD NOTE**
>
> This time we have just used '>0' or '<0' in the table. That's all you need.

> **REVISION NOTE**
>
> Check this calculation.

> **EXAM NOTE**
>
> Notice the 'error' term here. This is one-half of the interval length.

Interval bisection method

An alternative approach to improving a first interval is to check the sign of $y = f(x)$ at each end of the interval, and at the midpoint. You can then spot which half-sized interval contains the root. The process can be continued to an appropriate degree of accuracy.

EXAMPLE

Find the smallest solution of

$$x^3 - 3x + 1 = 0$$

A table provides a simple way of setting out the method:

	Interval		Midpoint	Sign of			m replaces	
n	a	b	m	$f(a)$	$f(b)$	$f(m)$	a	b
1	-2	-1	-1.5	negative	positive	positive		yes
2	-2	-1.5	-1.75	negative	positive	positive		yes
3	-2	-1.75	-1.875	negative	positive	positive		yes
4	-2	-1.875	-1.9375	negative	positive	negative	yes	
5	-1.9375	-1.875	etc.					

So the root lies in the interval $-1.9375 < x < -1.8750$, and so the best estimate of the root is the midpoint of this interval,

$$x = -1.906\,25 \pm 0.301\,25 \longleftarrow$$

You could find the remaining two roots of the equation by starting with a different interval.

> **METHOD NOTE**
>
> In the first row, $f(-2)$ is negative, $f(-1)$ is positive. At the midpoint $x = -1.5$ and then $f(-1.5)$ is also positive. So the graph must cross in the interval $-2 < x < -1.5$. And so on…

> **NOTE**
>
> The 'error' term is half the interval length.

Using the change of sign methods

Unless it is defined in a question, the choice of method is a matter of personal preference. For a typical root, the bisection method is likely to require fewer calculations than the decimal search. But in an exam, being organised is the most important thing.

Fixed-point methods

There is another approach to numerical methods of solution, starting from a single estimate of the solution instead of an interval. These approaches are known as **fixed-point methods**. Two common ones are formula iteration (see below) and Newton–Raphson iteration (see page 72).

Solving equations by formula iteration

Formula iteration depends on rearranging the equation to be solved into a form $x = f(x)$. An initial estimate of the solution can then be substituted in the right-hand side of this expression to find a new estimate. Another **iteration** can then be carried out by repeating the process with this new estimate. When this process is repeated again and again, successive approximations will either **converge** to a solution, or they will **diverge**.

It is easiest to see this using an example.

METHOD NOTE

This method is best done on a computer, using a spreadsheet.

In an exam you will have to make do with your calculator.

METHOD NOTE

There are many possible different formulae, including:

(1) $\quad x_{n+1} = \dfrac{(x_n)^3 + 1}{3}$

(2) $\quad x_{n+1} = \dfrac{3x_n - 1}{(x_n)^2}$

(3) $\quad x_{n+1} = \sqrt{\dfrac{3x_n - 1}{x_n}}$

(4) $\quad x_{n+1} = \dfrac{1}{3 - (x_n)^2}$

(5) $\quad x_{n+1} = \sqrt{3 - \dfrac{1}{x_n}}$

(6) $\quad x_{n+1} = \sqrt[3]{3x_n - 1}$

Check that you can get them all from the equation.

REVISION NOTE

By choosing suitable initial values for x_0 you can obtain all three solutions of the equation from the formulae above.

Try using various values to check the outcomes. Which diverge? Which converge? To which roots?

(The values of the roots are: $-1.879\,385\ldots$, $0.347\,296\ldots$, $1.532\,089\ldots$)

EXAMPLE

Find the smallest positive root of $x^3 - 3x + 1 = 0$.

The first task is to rearrange the equation into a form

$$x_{n+1} = f(x_n)$$

$$x^3 - 3x + 1 = 0 \implies 3x = x^3 + 1$$

$$\implies x = \frac{x^3 + 1}{3}$$

So we can use the formula

$$x_{n+1} = \frac{x_n{}^3 + 1}{3}$$

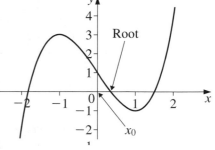

From the graph a reasonable first guess is $x_0 = 0$, and then successive values obtained from the formula are $0.333\,33\ldots$, $0.345\,68\ldots$, $0.347\,10\ldots$, $0.347\,27\ldots$, $0.347\,29\ldots$, and the sequence quickly **converges** to $0.347\,296\,355\ldots$

But this doesn't mean that the method always works! If we use the same formula and try to find the smallest root, starting with a reasonable guess of $x_0 = -2$, we get $-2.333\,33$, $-3.901\,23$, $-19.458\,4$, -2455.5, -4.9×10^9, ... and the sequence quickly **diverges** to negative infinity.

When you look for solutions by formula iteration, you should:

- Find several alternative iteration formulae (but exam questions often tell you which to use).
- Try each with different starting values to check for convergence or divergence.
- If possible, use a spreadsheet on a computer.

Why does formula iteration work?

For an iterative formula $x = f(x)$, because the method calculates a value of $f(x)$ and then uses the result to calculate again, the method can be viewed as a sequence of steps on a diagram:

1 Start from the initial estimate of the root on the x-axis
2 Draw a line up or down to the graph of $y = f(x)$
3 Draw a line across to $y = x$
4 Draw a line up or down to $y = f(x)$
5 Draw a line across to $y = x$
6 And so on, until the sequence either converges or diverges.

The diagrams show the method working or failing for different iteration formulae.

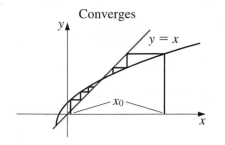
Converges

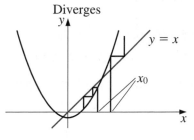
Diverges

REVISION NOTE

For obvious reasons, these diagrams are often called 'staircase' or 'cobweb' diagrams.

IMPORTANT

These diagrams show $y = f(x)$ and $y = x$ where the expression $f(x)$ is the right-hand side of the iteration formula $x_{n+1} = f(x_n)$. The solution is the x-value at the intersection of the two graphs.

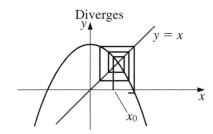
Diverges

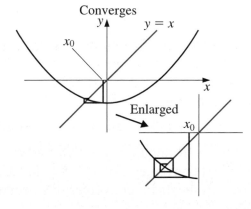
Converges

Enlarged

Inspection of these diagrams suggests that the iteration diverges if the gradient of $y = f(x)$ at the point where it crosses $y = x$ is not between -1 and $+1$.

It seems that the quickest possible convergence will happen if the graph of $y = f(x)$ has gradient zero at the crossing point.

The Newton–Raphson method

To solve the equation $g(x) = 0$, the Newton–Raphson formula is given by

$$x_{n+1} = x_n - \frac{g(x_n)}{g'(x_n)}$$

REVISION NOTE

This result is easy to prove, but not necessary for AS-Level.

The Newton–Raphson method is another formula iteration rearrangement of the equation, but a very special one with exactly the gradient property necessary to guarantee the 'quickest' convergence, as described above.

However, because it involves calculus, it is not always the easiest to use! Unless the method is defined, in an exam you can choose your own preferred method.

EXAMPLE

Use the Newton–Raphson method to solve the equation

$$x^3 - 3x + 1 = 0$$

Use $x_0 = 0$ as the initial approximation, and find two better estimates of a solution.

Put $f(x) = x^3 - 3x + 1$

$\Rightarrow f'(x) = 3x^2 - 3$

The N–R formula is

$$x_{n+1} = x_n - \frac{f(x_n)}{f'(x_n)}$$

$$= x_n - \frac{x_n^3 - 3x_n + 1}{3x_n^2 - 3}$$

$$= \frac{3x_n^3 - 3x_n - x_n^3 + 3x_n - 1}{3x_n^2 - 3}$$

$$= \frac{2x_n^3 - 1}{3x_n^2 - 3}$$

So if $x_0 = 0$, then $x_1 = 0.333\,333$

$$\Rightarrow x_2 = 0.347\,222$$

It you continue the N–R iteration, you will find that after just one more iteration it converges to $0.347\,296$. In fact, if you use a computer you will see that it converges to 9 decimal places in just 3 iterations.

Numerical integration – the trapezium rule

Numerical methods of finding an approximation to a definite integral are useful, for example, to find the area between a graph and the x-axis when the function for the graph cannot be easily integrated.

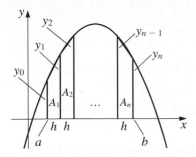

This method depends on dividing the required area into a number of equal width intervals, and approximating the area in each interval by a trapezium formed by joining the points on the graph at each end of the interval.

In the diagram, the area under the graph between $x = a$ and $x = b$ is divided into equal strips, each of width h.

Then, the areas of the trapezia can be approximated separately. For example, area A_2 is given by

$$A_2 = \tfrac{1}{2} \times h \times (y_1 + y_2)$$

and by summing all the separate trapezia, we get the

Trapezium rule for approximate integration

$$\int_a^b f(x)\,dx \approx \tfrac{1}{2}h\,[y_0 + 2(y_1 + y_2 + \ldots + y_{n-1}) + y_n]$$

where the interval from a to b is divided into n strips each of width h, and $y_0, y_1, y_2, \ldots, y_n$ are given by the values of $y = f(x)$ at the sides of the strips.

NOTE

There are other methods of numerical integration. The best known is 'Simpson's rule', but you will not need to know it for AS-Level.

METHOD NOTE

The accuracy of the method depends on how well the sloping top edges of the trapezia 'fit' the curve.

So as long as the curve doesn't bend too sharply, the fit is usually quite good.

Usually, dividing the area into more, 'thinner', strips will also improve the approximation.

EXAMPLE

Find an approximation for the area between the graph of $y = \sin^2 x$ and the x-axis between 0 and $\dfrac{\pi}{2}$ by using the trapezium rule with 4 strips.

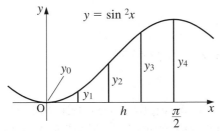

Each strip is of width $h = \dfrac{\pi}{8} \approx 0.393\ldots$

A table makes the calculation easy:

n	0	1	2	3	4
x_n	0.000	0.393	0.785	1.178	1.571
y_n	0.000	0.393	0.707	1.924	1.000

The trapezium rule gives the approximate area as

$$A \approx \tfrac{1}{2} \times (0.393) \times [0 + 2(0.383 + 0.707 + 0.924) + 1]$$

$$= 0.785 \text{ approximately.}$$

REVISION NOTE

The correct value of the area is
$$\int_0^{\frac{\pi}{4}} \sin^2 x\,dx = \frac{\pi}{4}$$

Numerical methods

1 For the equation

$$2x^3 + 3x^2 - x - 1.1 = 0$$

use the decimal search method to find the largest root of the equation, to 3 decimal places.

2 For the equation in question 1, use the method of interval bisection to find the smallest root.

3 For the equation in question 1, find a suitable formula iteration to obtain the middle root of the equation.

Check whether your iteration converges to the smallest or largest roots that you found in questions 1 and 2. If not, find alternative iterative formulae for each root, and check your answers to questions 1 and 2.

4 For the equation in question 1, find the Newton–Raphson iterative formula.

With suitable initial values, check that the Newton–Raphson formula converges to each of the three roots.

Check the speed of convergence of the solution using the Newton–Raphson formula with your solutions to questions 1, 2 and 3.

5 By using the trapezium rule with four intervals, find approximate values for:

(a) $\displaystyle\int_0^5 \frac{x}{1 + x^3}\, dx$ 　　　　　　(b) $\displaystyle\int_0^{\frac{\pi}{4}} \sqrt{\tan x}\, dx$

6 For the equation

$$x^3 - x^2 - x - 3 = 0$$

find the Newton–Raphson formula, and show that it can be written as

$$x_{n+1} = \frac{2x_n^3 - x_n^2 + 3}{3x_n^2 - 2x_n - 1}$$

Using $x_0 = 2$ as an initial approximation, find a better approximation.

7 (a) For the equation

$$2x^3 + 2x^2 + x - 1 = 0 \qquad\qquad (1)$$

show that the Newton–Raphson iteration formula is

$$x_{n+1} = \frac{4x_n^3 + 2x_n^2 + 1}{6x_n^2 + 4x_n + 1}$$

(b) If $y = 2x^3 + 2x^2 + x - 1$, find $\dfrac{dy}{dx}$.

By considering $\dfrac{dy}{dx} = 0$, show that there is only one root for equation (1).

(c) Find a suitable initial approximation x_0 for the root of the equation.

Use the Newton–Raphson iteration to find the root to 4 decimal places.

Answers and hints to solutions – Pure Maths

Exercise 1: Functions and mappings

1 (a) (i) $9x + 4$, range $\mathbb{R}$ (ii) $3x^2 + 1$, $\mathbb{R}: x \geq 1$
(iii) $9x^2 + 6x + 1$, $\mathbb{R}: x \geq 0$

(b) $f \circ f$ has an inverse, $h(x) = \dfrac{(x-4)}{9}$

2 (a)

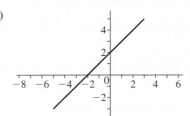

(b)

(c)

3 (a) (i) $(0, -\frac{3}{5})$, $(\frac{3}{4}, 0)$ (ii) $x = -\frac{5}{2}$ (iii) $y \to 2, y = 2$
(b)

$y = 2$

$x = -\frac{5}{2}$

4 (a)

(b) $(1, 1)$ (c)

$(1, 2)$

Exercise 2: Coordinate geometry

1 (a) $2y + x = 3$
(b) $5y + 4x = 6$, $l_3 : 4y - 5x = -29$

l_3 and l_2 cross at $(\dfrac{169}{41}, -\dfrac{86}{41})$, distance $\sqrt{\dfrac{81}{41}} \approx 1.41$

2 AC: $2y - x = 3$,
BC: $y - 4x = -29$, and AC
and BC cross at $\left(\dfrac{55}{7}, \dfrac{38}{7}\right)$,
so hypotenuse length ≈ 7.67

3 (a) M(15, 15) L(12, 3)
(b) $9x + 2y = 114$
(c) $3x - 5y = -30$
(d) $(10, 12)$

4 (a) $\sqrt{58}$ (b) $7x - 3y = 15$ (c) $\sqrt{116}$

5 (a) $2y = x + 1$, $2y = x + 8$ (b) B(5, 3) C(6, 7)
(c) $\sqrt{61}$, $\sqrt{13}$

Exercise 3: Algebra and series

1 (a) $-4\sqrt{3} < p < +4\sqrt{3}$ (b) $p = \pm 4\sqrt{3}$
(c) $p < -4\sqrt{3}$ or $p > +4\sqrt{3}$

2 $\dfrac{1 \pm \sqrt{19}}{4}$

3 (a) $k = 2$
(b) $2x^3 - x^2 + 2x - 16 = 0 \Rightarrow (x - 2)(2x^2 + 3x + 8) = 0$
for quadratic factor, discriminant < 0, so no real roots.

4 (a) $n = 20$ (b) $a = 16$, $r = \frac{1}{2}$

(c) (ii) $s_n = \dfrac{16\left(1 - \left(\frac{1}{2}\right)^n\right)}{1 - \frac{1}{2}} = 31.99 \Rightarrow 1 - \left(\frac{1}{2}\right)^n = 0.999\,687\,5$

$\Rightarrow \left(\frac{1}{2}\right)^n = 0.000\,312\,5 \Rightarrow n = 11.64$
so 12 terms needed

5 (a) *Scheme I*: £13 500   (b) *Scheme I*: £142 500
Scheme II: £13 498   *Scheme II*: £144 073

Exercise 4: Trigonometry

1 (a) $131.8°$, $-131.8°$
(b) $-163.8°$, $-136.2°$, $-43.8°$, $-16.2°$, $76.2°$, $103.8°$

2 2.61 km, 2.82 km

3 66.7 m

4 25.13 m, angle $= 0.57^c = 32.7°$

5 -2.84, 2.05

6 (a) $\dfrac{\sqrt{3} - 1}{2\sqrt{2}}$ (b) $2 \sin A \cos A$

7 $104.8°$, $97.9°$, $59.4°$, $97.9°$

Exercise 5: Calculus – understanding change

1 (a) $x^2(5x^2 - 6)$ (b) $-\dfrac{9}{x^4} - \dfrac{1}{3}x^{-\frac{2}{3}}$

(c) $\cos x + \sin x$

2 (a) $3y + 12x = 19$ (b) $\dfrac{361}{72}$ units^2

(c) $3y = 12x - 5$

3 (a) maximum at $(-5, \frac{13}{3})$ minimum at $(-1, -\frac{25}{3})$

(b) crosses y-axis at $(0, -4)$ $\dfrac{d^2y}{dx^2} = 0 \Rightarrow$ inflexion at $(-3, -1)$

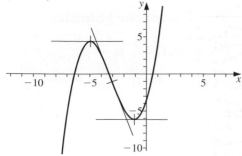

4 (a) $\frac{1}{4}x^4 + \frac{2}{3}x^3 + \frac{1}{2}x^2 + C$

(b) $\frac{3}{4}x^{\frac{4}{3}} - \dfrac{1}{x^3} + C$

(c) $\sin x - 2\cos x + C$

5 $81\frac{2}{3} = \frac{245}{3}$ units^2

6 $A_1 = \displaystyle\int_0^1 x^3 - 3x^2 + x + 1 \, dx = \frac{3}{4}$

$A_2 = \displaystyle\int_1^2 x^3 - 3x^2 + x + 1 \, dx = -\frac{3}{4}$,

Area $= A_1 - A_2 = 2$ units^2

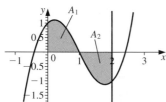

7 $x^2 - 1 = -x^2 + 5x - 4 \Rightarrow x = \frac{3}{2}$ or $x = 1$

$A_1 = \displaystyle\int_1^{\frac{3}{2}} x^2 - 1 \, dx = \frac{7}{24}$

$A_2 = \displaystyle\int_1^{\frac{3}{2}} -x^2 + 5x - 4 \, dx = \frac{1}{3}$

$\Rightarrow$ Area $= \frac{1}{24}$ units^2

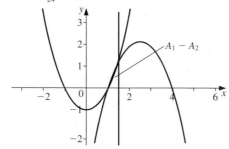

8 $V = \displaystyle\int_0^4 \pi(3 + \sqrt{x})^2 \, dx = 76\pi$ units^3

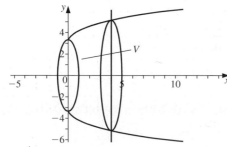

9 $V = \pi \displaystyle\int_0^4 y \, dy = 8\pi$

(Note that this question asks for rotation about the y-axis, so the method is $V = \pi \displaystyle\int_a^b x^2 \, dy$)

Exercise 6: Differentiation and integration

1 (a) $18x^2(x^3 + 2)^5$

(b) $\dfrac{3}{(x + 2)^2}$

(c) $4x^3 + 3x^2 + 1$

(d) $\dfrac{3x^2}{2\sqrt{x^3 + 1}}$

(e) $\cos x - x \sin x$

(f) $\dfrac{-(2x \sin x + \cos x)}{2x^{3/2}}$

(g) $xe^{-x}(2 - x)$

(h) $\dfrac{3x}{2\sqrt{x + 1}}$

2 $\dfrac{dr}{dt} = \dfrac{1}{18\pi} \approx 0.018 \text{ m s}^{-1} \approx 2 \text{ cm s}^{-1}$

3 (a) $\frac{1}{18}(x^3 + 1)^6 + C$

(b) $\frac{2}{3}(\sqrt{x^3 - 1}) + C$

4 (b) 18 m s^{-1} downwards

5 $-\frac{1}{3}\cos^3 x + k$

7 (a) $e^x(\cos x - \sin x)$

(b) $\dfrac{x(x - 2)}{(x - 1)^2}$

(c) $8x(x^2 - 3)^3$

(d) $\cos^2 x - \sin^2 x = \cos 2x$

8 (a) $\frac{1}{8}(x^2 - 3)^4 + C$

(b) $\dfrac{1}{\cos x} + C = \sec x + C$

(c) $\sqrt{2} - 1$

Exercise 7: Logs and exponentials

1 (a) 4.292

(b) 2

(c) 14.43

2 $2(\ln 8 - \ln 3) = 1.962$

3 (a) $xe^{-3x}(2 - 3x)$

(b) $\dfrac{2e^{2x}(x - 2)}{x^5}$

4 (b) $k = 5.21$

(c) $n = 0.325$, curve is $y \approx 5.21x^{0.325}$

5 (b)

x	1	2	3	4	5	6
y	2.21	6.39	8.76	11.67	18.41	20.44
$\log_{10} x$	0	0.301	0.477	0.602	0.699	0.778
$\log_{10} y$	0.344	0.806	0.943	1.067	1.265	1.310

(c) $A \approx 2.36$, $n \approx 1.23$

Exercise 8: Numerical methods

1 From the table the root is 0.638 approx.

h =	0.1		0.01		0.001		0.0001	
x	f(x)	x	f(x)	x	f(x)	x	f(x)
0	−1.1000	0.6	−0.1880	0.63	−0.0392	0.637	−0.0027
0.1	−1.1680	0.61	−0.1397	0.631	−0.0340	0.6371	−0.0022
0.2	−1.1640	0.62	−0.0901	0.632	−0.0289	0.6372	−0.0017
0.3	−1.0760	**0.63**	**−0.0392**	0.633	−0.0237	0.6373	−0.0012
0.4	−0.8920	**0.64**	**0.0131**	0.634	−0.0185	0.6374	−0.0006
0.5	−0.6000	0.65	0.0668	0.635	−0.0132	**0.6375**	**−0.0001**
0.6	**−0.1880**	0.66	0.1218	0.636	−0.0080	**0.6376**	**0.0004**
0.7	**0.3560**	0.67	0.1782	**0.637**	**−0.0027**	0.6377	0.0009
0.8	1.0440	0.68	0.2361	**0.638**	**0.0025**	0.6378	0.0015
0.9	1.8880	0.69	0.2953	0.639	0.0078	0.6379	0.0020
1	2.9000	0.7	0.3560	0.64	0.0131	0.638	0.0025

2 If the table is continued, the root can be shown to be −1.597 approximately.

n	a	b	m	f(a)	f(b)	f(m)	m replaces
0	−2.0000	−1.0000	−1.5000	−3.1000	0.9000	0.4000	b
1	−2.0000	−1.5000	−1.7500	−3.1000	0.4000	−0.8813	a
2	−1.7500	−1.5000	−1.6250	−0.8813	0.4000	−0.1352	a
3	−1.6250	−1.5000	−1.5625	−0.1352	0.4000	0.1573	b
4	−1.6250	−1.5625	−1.5938	−0.1352	0.1573	0.0175	b
5	−1.6250	−1.5938	−1.6094	−0.1352	0.0175	−0.0572	a
6	−1.6094	−1.5938	−1.6016	−0.0572	0.0175	−0.0195	a
7	−1.6016	−1.5938	−1.5977	−0.0195	0.0175	−0.0009	a
8	−1.5977	−1.5938	−1.5957	−0.0009	0.0175	0.0083	b
9	−1.5977	−1.5957	−1.5967	−0.0009	0.0083	0.0037	b
10	−1.5977	−1.5967	−1.5972	−0.0009	0.0037	0.0014	b

3 All of the following formulae can be obtained:

(1) $x = 2x^3 + 3x^2 - 1.1$

(2) $x = \dfrac{-2x^3 + x + 1.1}{3x}$

(3) $x = \sqrt{\left(\dfrac{-2x^3 + x + 1.1}{3}\right)}$

(4) $x = \dfrac{-3x^2 + x + 1.1}{2x^2}$

(5) $x = \sqrt{\dfrac{-3x^2 + x + 1.1}{2x}}$

(6) $x = \sqrt[3]{\dfrac{-3x^2 + x + 1.1}{2}}$

(7) $x = \dfrac{x + 1.1}{2x^2 + 3x}$

(8) $x = \sqrt{\dfrac{x + 4}{2x + 3}}$

(9) $x = \dfrac{1.1}{2x^2 + 3x - 1}$

They converge for suitable starting values as below:

	Formula	1	2	3	4	5	6	7	8	9
Converges to	Lowest root	no	no	no	yes	no	yes	no	no	no
	Middle root	no	no	no	no	no	no	no	no	yes
	Highest root	no	no	yes	no	no	no	yes	yes	no

4 N–R formula is

$$x_{n+1} = x_n - \frac{2x_n^3 + 3x_n^2 - x_n - 1.1}{6x_n^2 + 6x_n - 1}$$

It converges to all three roots as in the table on the right.

	Newton–Raphson Formula		
u_0	0.0	−1.0	−2.0
u_1	−1.1	−0.1	−1.718 18
u_2	1.747 059	−0.731 17	−1.613 55
u_3	1.136 374	−0.523 21	−1.597 82
u_4	0.799 321	−0.540 03	−1.597 47
u_5	0.663 162	−0.540 05	−1.597 47
u_6	0.638 332	−0.540 05	−1.597 47
u_7	0.637 522	−0.540 05	−1.597 47

5 (a) $h = 1.25$

n	0	1	2	3	4
x_n	0	1.25	2.5	3.75	5
y_n	0	0.4233	0.1504	0.0698	0.0397

$A = 0.8291$

(b) $h = 0.1963$

n	0	1	2	3	4
x_n	0	0.1963	0.3927	0.589	0.7854
y_n	0	0.466	0.6436	0.8174	1

$A = 0.4726$

6 2.142 857 143

7 (b) $\dfrac{dy}{dx} = 6x^2 + 4x + 1$, $\dfrac{dy}{dx} = 0$ has no real solutions, so the cubic has no stationary points.

Hence its graph can cross the x-axis only once, so there is only one root for equation (1).

(c) 0.4406

Statistics Topics:

1 Discrete data

Measures of centre 79

Measures of spread 79

Quartiles 80

Stem and leaf diagrams 80

Box plots (box and whisker plots) 80

Interquartile range 80

Exercise 1: Discrete data 81

2 Continuous data

Measures of centre 82

Measures of spread 82

Histograms 83

Properties of standard deviation 83

Cumulative frequency graphs 84

Coding 85

Exercise 2: Continuous data 86

3 Orderings

Arrangements where order is important 87

Arrangements where order is not important 87

Sampling with replacement 88

Sampling without replacement 88

Exercise 3: Orderings 89

4 Probability

Venn diagrams 90

Mutually exclusive events 91

Independence 91

Combined events 92

Conditional probability 92

Exercise 4: Probability 93

5 Discrete random variables

Definition 94

Probability density functions 94

The cumulative distribution function 95

Expectation of a discrete random variable 95

Variance of a discrete random variable 95

Exercise 5: Discrete random variables 96

6 Discrete probability distributions

The binomial distribution 97

Hypothesis testing on a binomial distribution 99

The Poisson distribution 101

Discrete uniform distribution 103

Geometric distribution 103

Exercise 6: Discrete probability distributions 104

7 Expectation algebra

Definitions 105

Expectations of functions of random variables 105

More formulae I 105

More formulae II 106

Exercise 7: Expectation algebra 107

8 Continuous random variables

Definition 108

The normal distribution 108

The standard normal distribution 109

Other normal distributions 110

Finding μ or σ from given information 110

Exercise 8: Continuous random variables 111

9 Sampling

Why consider a sample? 112

Random sampling methods 112

Non-random sampling methods 112

10 Correlation and regression

Pearson's product moment correlation coefficient 114

Linear regression 115

Exercise 10: Correlation and regression 116

Answers and hints to solutions 117

1 Discrete data

Discrete data can only take a finite number of fixed values, for example, the shoe sizes of a group of 6-year-old children, or the number of goals scored by a football team in games played during one season.

REVISION NOTE

Discrete data can be counted.

Measures of centre

1   The **mean** $\bar{x} = \dfrac{\Sigma fx}{\Sigma f}$ or $\dfrac{\Sigma x}{n}$ (the arithmetic average)
    $\bar{x}$ denotes the mean of a sample.

2   The **mode** is the value that occurs most often.

3   The **median** is the middle value when the data is arranged in ascending order.

REVISION NOTE

Mode is not used for small data sets.

REVISION NOTE

Median is best for skewed data.

Measures of spread

1   The **range** is the difference between the highest and the lowest entries in the set.

2   The **variance**, $s^2 = \dfrac{1}{n}\Sigma fx^2 - \bar{x}^2$.

3   The **standard deviation**, $s = \sqrt{\dfrac{1}{n}\Sigma fx^2 - \bar{x}^2}$
    s denotes the standard deviation of a sample.

REVISION NOTE

This may also be written as

$$s = \sqrt{\dfrac{\Sigma f(x-\bar{x})^2}{n}} \text{ or } \sqrt{\dfrac{\Sigma x^2 - x^2}{n}}$$

You must memorise this formula.

EXAMPLE

The following scores were recorded when 50 students sat a multiple choice test consisting of 10 questions.

x	3	4	5	6	7	8	9	10
f	4	5	2	12	7	5	6	9

Calculate the mean and standard deviation of the scores.

The solution to this problem could be set out using a table, but it is quicker to do the calculations on your calculator. Input the data with the calculator in single variable mode.

Your calculator will give you the following summary statistics:

$$\Sigma x = \Sigma fx = 347, \ n = 50 \ \text{ and } \ \Sigma x^2 = \Sigma fx^2 = 2647$$

$$\bar{x} = \frac{\Sigma fx}{\Sigma f} = \frac{347}{50} = 6.94$$

$$s = \sqrt{\frac{\Sigma fx^2}{\Sigma f} - \bar{x}^2}$$

$$= \sqrt{\frac{2647}{50} - 6.94^2}$$

$$= \sqrt{4.7764} = 2.185$$

Mean $= 6.94$ and standard deviation $= 2.185$

METHOD NOTE

A data set is often easier to understand and analyse when it is displayed in a frequency table.

EXAM NOTE

Make sure that you show enough working out to explain your answers.

REVISION NOTE

The smaller the standard deviation, the more consistent the data.

Quartiles

For small discrete data sets:
- the **lower quartile** is at the $\frac{1}{4}(n+1)$ position
- the **median** is at the $\frac{1}{2}(n+1)$ position
- the **upper quartile** is at the $\frac{3}{4}(n+1)$ position.

EXAMPLE

Listed below are the numbers of words in 27 sentences selected at random from a popular novel.

4, 32, 15, 4, 20, 5, 34, 6, 6, 36, 6, 15, 7, 17, 36, 8, 26, 24, 8, 18, 10, 19, 19, 12, 23, 25

Find the median sentence length and the quartiles.

Lower quartile is at the $\frac{1}{4}(27+1) = $ 7th position.

Median is at the $\frac{1}{2}(27+1) = $ 14th position.

Upper quartile is at the $\frac{3}{4}(27+1) = $ 21st position.

Order the data then simply count along the row to the 7th, 14th and 21st positions to find the values required.

4, 4, 5, 6, 6, 6, (7), 8, 8, 10, 12, 12, 15, (15,) 17, 18, 19, 19, 20, 23, (24,) 25, 26, 32, 34, 36, 36

Median is 15. Lower quartile is 7. Upper quartile is 24.

Stem and leaf diagrams

The lengths of the 27 sentences in the example above can be ordered using a stem and leaf diagram. The first digit of each data point becomes the stem and the following digits become the leaves.

```
0 | 4  4  5  6  6  6  (7)  8  8
1 | 0  2  2  5  (5)  7  8  9  9
2 | 0  3  (4)  5  6
3 | 2  4  6  6
```

$n = 27$ and $2|4 = 24$

Box plots (box and whisker plots)

A box plot gives a more pictorial view of the same data. All the important values can be seen easily.

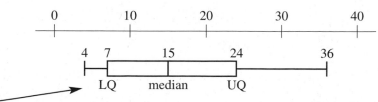

Interquartile range

The **interquartile range** is the difference between the upper and lower quartiles, or UQ − LQ.

For the data in the example, IQR $= 24 - 7 = 17$.

This shows that the middle 50% of the data ranges over 17 values.

Discrete data

1 (*a*) Find the median, the lower and upper quartiles and the interquartile range for the following temperatures in °C recorded over a two-week period during the summer.

| 34 | 25 | 33 | 37 | 20 | 31 | 38 |
| 21 | 26 | 31 | 31 | 27 | 23 | 32 |

(*b*) Find the median, the upper and lower quartiles and the interquartile range for the following temperatures in °C recorded over a two-week period in the winter.

| 17 | 17 | 14 | 16 | 15 | 10 | 14 |
| 4 | 6 | 3 | 10 | 6 | 2 | 8 |

(*c*) Using the same scale for each, draw a box plot to represent each set of data and comment on any similarities and differences.

2 The following test results were recorded for two different classes of maths students completing the same test.

Class A

| 87 | 99 | 72 | 50 | 77 | 59 | 90 | 73 | 67 | 45 | 78 |
| 84 | 70 | 77 | 79 | 88 | 47 | 66 | 80 | 96 | 55 | 70 |

Class B

| 58 | 82 | 56 | 58 | 50 | 57 | 65 | 27 | 45 | 53 |
| 83 | 62 | 73 | 79 | 77 | 64 | 48 | 53 | 48 | 42 |

Calculate the mean and standard deviation for both classes. Comment on any similarities and differences.

3 (*a*) The following data on absenteeism was collected from a small factory over a period of 25 days.

$$\Sigma x = 40 \quad \Sigma x^2 = 126$$

Calculate the mean and standard deviation for this 25-day period.

(*b*) For the 25-day period immediately following, the mean number of workers absent was 1.2 and the standard deviation was 1.6.

Calculate the combined mean and standard deviation for the whole 50 days.

4 Draw an ordered stem and leaf plot for this data.

| 23 | 25 | 31 | 39 | 26 | 47 | 53 | 28 | 46 | 45 | 46 |
| 50 | 41 | 40 | 46 | 20 | 47 | 22 | 43 | 38 | 55 | 24 |

Mark the median, upper and lower quartiles on it and calculate the interquartile range.

Cumulative frequency graphs

These are used to find the median, the quartiles and the interquartile range for grouped data. Plot the running total (cumulative frequency) against the upper class boundary to get a smooth increasing curve.

40 students were asked to hold their breath for as long as possible and the times, in seconds, were recorded below. Use this information to draw a cumulative frequency graph, then find the median and the quartiles.

Class	$0 \leqslant t < 15$	$15 \leqslant t < 30$	$30 \leqslant t < 45$	$45 \leqslant t < 60$	$60 \leqslant t < 90$	$90 \leqslant t < 120$
f	5	8	9	8	6	4

Cumulative frequencies:

	<15	<30	<45	<60	<90	<120
C.f.	5	13	22	30	36	40

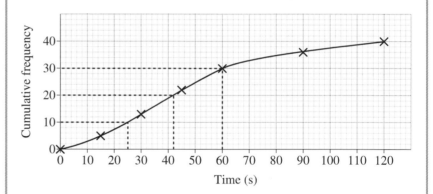

METHOD NOTE

Add a cumulative frequency column or row to your table. Use the upper class boundaries to indicate where to plot the running total.

EXAM NOTE

Draw a smooth curve through the points.

EXAM NOTE

Be careful here. The graph should start at zero at the lower boundary of the first class. The cumulative frequencies must be plotted against the upper boundary of each class.

REMEMBER

The middle 50% of data is in the IQR.

Because the data set is large (i.e. 40 or more) the median is at the $\frac{1}{2}n$ position and the quartiles are at the $\frac{1}{4}n$ and $\frac{3}{4}n$ positions.

The **median** is at cumulative frequency 20. Read across to the graph and down to the time axis. Median = 42

The **lower quartile** is at cumulative frequency 10. Read across to the graph and down to the time axis. Lower quartile = 25

The **upper quartile** is at frequency 30. Read across to the graph and down to the time axis. Upper quartile = 60

Interquartile range = Upper quartile − Lower quartile
$$= 60 - 25 = 35$$

Semi-interquartile range = $\frac{1}{2}$ interquartile range
$$= 17.5$$

These two measures of spread are useful when comparing skewed data sets. The smaller the IQR, the more consistent the data.

Coding

Data can be coded to make calculations simpler.

> If $y = ax + b$
>
> is used to modify a data set for ease of calculation, then
>
> $\bar{y} = a\bar{x} + b$
>
> where $\bar{y}$ and $\bar{x}$ are the corresponding means and
>
> $s_y = |a|s_x$

REVISION NOTE

The value of the mean is changed by both the multiplier and the constant.

REVISION NOTE

Standard deviation is always positive and is only changed by the multiplier.

EXAMPLE

The heights of 6 daffodils, measured to the nearest centimetre, were:

    30,   33,   37,   28,   29,   26

Use the code $y = x - 30$ to calculate the mean and standard deviation for the heights.

EXAM NOTE

Use a code that gives the simplest possible calculations.

Set out your calculations in a table.

x	y	y^2
26	−4	16
28	−2	4
29	−1	1
30	0	0
33	3	9
37	7	49
Σ	3	79

$\bar{y} = \frac{3}{6} = 0.5$

$s_y = \sqrt{\frac{79}{6} - 0.5^2} = 3.59$

Since $y = x - 30$

    $\bar{y} = \bar{x} - 30 \Rightarrow \bar{x} = \bar{y} + 30$

    $\bar{x} = 30.5$

and $s_y = 1 \times s_x \Rightarrow s_x = 3.59$

METHOD NOTE

You can check this using the original raw data:

$\Sigma x = 183, \quad \Sigma x^2 = 5659$

$\Rightarrow \bar{x} = \dfrac{183}{6} = 30.5$

$s_x = \sqrt{\dfrac{5659}{6} - 30.5^2}$

$= 3.59$

EXAMPLE

In an angling competition, anglers calculate the mean and standard deviation of the lengths of their best five fish. One angler calculates the mean to be 17.2 inches and the standard deviation to be 1.5 inches.

The angler should have used metric measures.

Calculate the metric mean and standard deviation using the code 1 inch = 2.5 cm.

Code is $y = 2.5x$

So $\text{mean}_{\text{metric}} = 2.5 \times \text{mean}_{\text{imperial}}$

    $\text{mean}_{\text{metric}} = 2.5 \times 17.2 = 43 \text{ cm}$

      $s_m = 2.5 s_{\text{imp}} = 2.5 \times 1.5 = 3.75 \text{ cm}$

REVISION NOTE

Once the raw data is lost, coding is the only way to get the correct measurements.

Continuous data

1 The table shows the time (in minutes) spent travelling to work for 100 people.

Time	0–9	10–19	20–29	30–49	50–89
Frequency	15	35	37	8	5

(*a*) Write down the actual class intervals and class width for each class.

(*b*) Calculate the frequency density for each class and draw a histogram to illustrate the data set.

(*c*) Calculate the mean and the standard deviation.

2 The following histogram illustrates data collected about the height of sunflower plants, four weeks after planting the seeds.

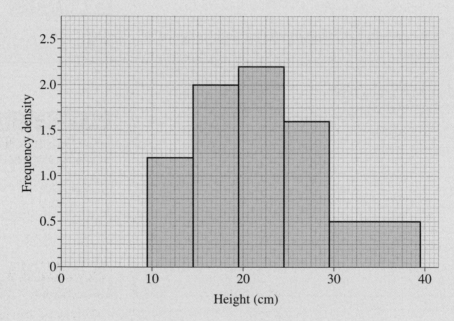

(*a*) If 40 plants were measured, how many plants were in the classes 10–14 and 25–29?

(*b*) Draw up a cumulative frequency table for this data and complete a cumulative frequency graph.

(*c*) Find the median and the interquartile range from the graph.

3 Over a period of 30 consecutive days, the average time to complete a bus journey on a particular route was 1.2 hours and the standard deviation of the times was 0.07 hours.

What was the average journey time and standard deviation in minutes?

4 Find the mean and standard deviation for the data set below, without using the statistical functions on your calculator. Use a suitable code to simplify the calculations.

x	11	21	31	41	51
f	3	5	11	7	4

3 Orderings

Arrangements where order is important

Here the order of choosing is important. For example, order is important to the athletes who gain 1st, 2nd and 3rd in a race.

Order is important when objects are arranged in a line.

The number of ways n objects can be arranged in a line.

$n!$

The number of ways n objects can be arranged in a line, when a of these objects are of one type, b are of a second type, c are of a third type, etc.

$$\frac{n!}{a!b!c!\ldots}$$

METHOD NOTE

4 people:
$4! = 4 \times 3 \times 2 \times 1$ ways of arranging them in a line.

The number of different ways of filling r places from a group of n objects.

$$^{n}P_{r} = \frac{n!}{(n - r)!}$$

These are called **permutations**.

EXAMPLE

How many different ways can the letters in the word 'hippopotamus' be arranged?

'Hippopotamus' contains 12 letters, including 3p's and 2o's.

$$\frac{12!}{2!3!} = 39\,916\,800$$

EXAM NOTE

You must show this much working out to get all your marks.

EXAMPLE

There are 8 people and 5 chairs. How many different ways can you fill the 5 chairs?

$n = 8, \quad r = 5, \quad$ so

$$^{8}P_{5} = \frac{8!}{(8 - 5)!}$$
$$= 6720$$

EXAM NOTE

Ask yourself:
'Does the order matter?'

Arrangements where order is not important

Here the content of the group is important, but the order they are chosen in is irrelevant.

The number of ways of selecting a group of r objects from a larger group of n objects.

$$^{n}C_{r} = \frac{n!}{(n - r)!r!}$$

These are called **combinations**.

EXAMPLE

How many groups of 5 people can be chosen from a total of 8 people?

$n = 8, \quad r = 5, \quad$ so

$$^{8}C_{5} = \frac{8!}{(8 - 5)!5!}$$
$$= 56$$

REVISION NOTE

You need to learn how to use the $^{n}P_{r}$ and $^{n}C_{r}$ functions on your calculator.

Sampling with replacement

The probability of a particular event does not change during the experiment, because all items are available for selection each time.

Sampling without replacement

In this case a group of items is chosen all at once. For example, you put your hand into a bag and pull out 3 marbles. The order in which they arrive in your hand is not important. It is the make-up of the group that is important.

METHOD NOTE

In this case order is important.

METHOD NOTE

You must make sure you include all possible arrangements of one B and two B.

Order is important.

METHOD NOTE

In this case order isn't important.

REVISION NOTE

This problem is similar to choosing a committee made up of three people, where one is male and two are female.

Orderings

1   Find the probability that a group of two girls and one boy is chosen at random to represent their class at a presentation. A class consists of 12 boys and 15 girls.

2   Postcodes consist of two letters and two digits, together with one digit and two letters.

   How many possible different postcodes are there if all 26 letters may be used with all digits 0–9? All letters and digits may be repeated.

3   How many different arrangements can be made of the letters in the words:

   (*a*) oranges,

   (*b*) apples,

   (*c*) bananas?

4   How many ways can you choose at random a group of three videos from a box of ten?

5   A bag contains 4 oranges, 3 apples and 2 pears. Three pieces of fruit are chosen at random without replacement.

   Find the probability that:

   (*a*) no oranges are chosen,

   (*b*) 3 oranges are chosen,

   (*c*) exactly one plum is chosen.

6   A family consists of a mother, father, 3 sons and 4 daughters. They win 4 tickets to the cinema.

   Calculate the number of ways in which a group of four could be chosen from this family if it must contain:

   (*a*) any four members of the family,

   (*b*) both parents,

   (*c*) exactly one parent,

   (*d*) neither parent.

7   The family in question 6 decides to choose the group by drawing names out of a hat. Calculate the probability that:

   (*a*) the group is all female,

   (*b*) the group is all male,

   (*c*) the group consists of the mother and the 3 sons.

4 Probability

All probabilities must take values between 0 and 1.

EXAM NOTE

Drawing a Venn diagram will make calculations of some probabilities easier.

Venn diagrams

A Venn diagram shows all possible outcomes of an experiment and any events of interest. For example, when two 4-sided dice are rolled there are 16 possible outcomes. (Each die is numbered 1 to 4.)

Let event A = {at least one 4} and

event B = {total is five}

The Venn diagrams below show all possible outcomes, with event A and event B shaded.

(1, 1) (2, 1)	**A**
(1, 2) (2, 2)	(1, 4) (2, 4)
(1, 3) (2, 3)	(4, 1) (4, 4) (4, 2)
(3, 3) (3, 2)	(3, 4) (4, 3)
(3, 1)	

(1, 1) (2, 1) (3, 1)	**B**
(1, 2) (2, 2) (4, 2)	(1, 4)
(1, 3) (3, 3) (3, 4)	(2, 3) (4, 1)
(4, 4) (2, 4) (4, 3)	(3, 2)

You can see that $P(A) = \frac{7}{16}$ and $P(B) = \frac{4}{16}$.

Common terms

Complement
The event A′ means 'not A' and B′ means 'not B'.

$$P(A') = 1 - P(A)$$

A′	
(1, 1) (2, 1)	**A**
(1, 2) (2, 2)	(1, 4) (2, 4)
(1, 3) (2, 3)	(4, 1) (4, 4) (4, 2)
(3, 3) (3, 2)	(3, 4) (4, 3)
(3, 1)	

B′	
(1, 1) (2, 1) (3, 1)	**B**
(1, 2) (2, 2) (4, 2)	(1, 4)
(1, 3) (3, 3) (3, 4)	(2, 3) (4, 1)
(4, 4) (2, 4) (4, 3)	(3, 2)

Union
A ∪ B

Objects belong to either A or B or both A and B.

Here $P(A \cup B) = \frac{9}{16}$

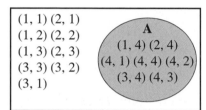

Intersection
A ∩ B

Objects must belong to both A and B.

Here $P(A \cap B) = \frac{2}{16}$

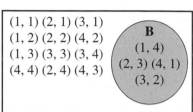

Venn diagrams (cont.)

An important formula:

$$P(A \cup B) = P(A) + P(B) - P(A \cap B)$$
$$= \tfrac{7}{16} + \tfrac{4}{16} - \tfrac{2}{16} \quad \text{(in the Venn diagrams on page 90)}$$
$$= \tfrac{9}{16}$$

REVISION NOTE

Learn this formula.

$P(A \cap B)$ has been counted twice: once with $P(A)$ and again with $P(B)$.

Mutually exclusive events

Mutually exclusive events have nothing in common.

EXAMPLE

The Venn diagram below lists all the possible outcomes when two 6-sided dice are rolled.

Let event C = {the sum of the two dice is even}
Let event D = {the sum of the two dice is odd}

Show that C and D are mutually exclusive.

C	**D**
(1, 1)	(1, 2)
(1, 3) (1, 5) (2, 2)	(1, 4) (1, 6) (2, 1)
(2, 4) (2, 6) (3, 1)	(2, 3) (2, 5) (3, 2)
(3, 3) (3, 5) (4, 2)	(3, 4) (3, 6) (4, 1)
(4, 4) (4, 6) (5, 1)	(4, 3) (4, 5) (5, 2)
(5, 3) (5, 5) (6, 2)	(5, 4) (5, 6) (6, 1)
(6, 4) (6, 6)	(6, 3) (6, 5)

$P(C) = \tfrac{18}{36}$
$P(D) = \tfrac{18}{36}$

Events C and D have no outcomes in common, so C and D are mutually exclusive.

EXAM NOTE

To prove that two events are mutually exclusive, show that their intersection = 0.

For mutually exclusive events:

$$P(C \cap D) = 0 \text{ and } P(C \cup D) = P(C) + P(D)$$

REVISION NOTE

This is a modification of the important formula above.

Independence

If two events F and G are **independent**, then the occurrence of F cannot influence the occurrence of G, and vice versa. For example, a tossed coin lands with tails up on the first toss. When the coin is tossed again, the fact that it landed tails up the first time will not influence the way that it lands the second time. Both coin tosses are independent of each other.

For independent events:

$$P(F \cap G) = P(F) \times P(G)$$

REVISION NOTE

This is not the same as mutually exclusive, because independent events can occur together.

Events C and D from the example above are not independent, because

$$P(C \cap D) = 0 \text{ but } P(C) \times P(D) = \tfrac{18}{36} \times \tfrac{18}{36} = \tfrac{1}{4}$$

You must never use this formula to calculate the probability of two events happening together (i.e. $P(F \cap G)$) *unless* you are told that the events are independent. This is particularly important when considering conditional probability.

Combined events

Here we are looking for the probability of two or more things happening together.

EXAMPLE

A fair 4-sided die is rolled, then a fair coin is tossed. Find the probability that the number is even and the coin toss results in a head.

Draw a **tree diagram** to show all possible outcomes.

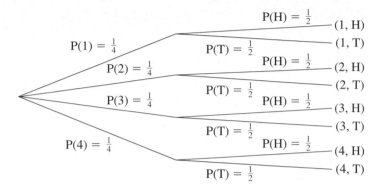

From the diagram, the outcomes required are (2, H) and (4, H).

$$P(2, H) + P(4, H) = \left(\tfrac{1}{4} \times \tfrac{1}{2}\right) + \left(\tfrac{1}{4} \times \tfrac{1}{2}\right) = \tfrac{2}{8}$$

Conditional probability

Suppose there are two events A and B that could occur during an experiment. For example, A is the event 'choosing a heart from a pack of cards', while B is the event 'choosing an even card (not a picture)'.

If we know that event A has occurred, this will change the way we calculate the probability that B may occur.

P(B|A) means 'the probability that event B occurs, given that event A has already occurred'.

This can be calculated using the formula:

$$P(B|A) = \frac{P(A \cap B)}{P(A)}$$

EXAMPLE

Two cards are chosen from a pack of 52. Find the probability of choosing an even card as the second card, given that the first card was a heart.

$P(A, \text{choosing a heart}) = \tfrac{13}{52}$ $P(B, \text{an even card}) = \tfrac{20}{52}$

$P(A \text{ and } B) = P(A \cap B) = \tfrac{5}{52} = P(\text{even and heart})$

$P(B|A) = \dfrac{P(A \cap B)}{P(A)} = \dfrac{\frac{5}{52}}{\frac{13}{52}} = \tfrac{5}{13}$

If A and B are **independent** $P(B|A) = P(B)$

If A and B are **mutually exclusive** $P(B|A) = 0$

Probability

1   A and B are two events, where

   $P(A) = \frac{3}{7}$, $P(B) = \frac{1}{5}$ and $P(A \cup B) = \frac{19}{35}$

   (*a*) Find $P(A \cap B)$

   (*b*) Are A and B independent?

   (*c*) Find $P(A')$ and $P(B')$.

HINT

You will need the formula from page 91 here.

2   C and D are two events, where

   $P(C) = \frac{2}{9}$, $P(D) = \frac{5}{12}$ and $P(C \cup D) = \frac{23}{36}$

   (*a*) Find $P(C \cap D)$

   (*b*) Are C and D independent?

   (*c*) Are C and D mutually exclusive?

3   A number is drawn at random from a bag containing the numbers 1–12.

   Given that the number is odd, what is the probability that it is a multiple of 3?

HINT

Write the formula out first, then write out in words what you actually need to find.

4   A and B are events such that $P(A) = \frac{1}{3}$, $P(B) = \frac{5}{12}$

   $P(A|B) = \frac{1}{6}$.

   Find $P(A \cap B)$ and $P(B|A)$.

   Are A and B independent?

5   A bag contains 25 packets of crisps, of which 20 are ready salted and 5 are cheese and onion. One of the cheese and onion packets contains a £5 note, as do three of the ready salted packets. Two packets of crisps are chosen from the bag.

   Find:

HINT

A tree diagram will make this question easier.

   (*a*) The probability that both packets of crisps contain a £5 note.

   (*b*) The probability that both packets of crisps are ready salted and at least one contains a £5 note.

   (*c*) The probability that at least one packet contains a £5 note, given that both packets are ready salted.

   (*d*) The probability that both packets are ready salted given that at least one packet is ready salted.

6   In Beachtown, each day is either fine or wet. If it is fine one day, the probability that the next day is also fine is 0.8. If it is wet one day, the probability that it is wet the next day is 0.6.

   Given that Thursday is fine; find:

HINT

A tree diagram will make this question easier.

   (*a*) The probability that Friday is wet.

   (*b*) The probability that Friday is wet and Saturday is fine.

   (*c*) The probability that Saturday is fine.

5 Discrete random variables

Definition

For X to be a discrete random variable it must have the following qualities.

1   The values of X are discrete and variable.

2   X can only assume certain values, $x_1, x_2, ..., x_n$.

3   Each value of X has its own probability, $p_1, p_2, ..., p_n$ such that $p_1 + p_2 + ... + p_n = 1$

EXAMPLE

X is the discrete random variable defined as 'the sum of the scores shown by two 4-sided dice'.

Display the probability distribution in a table.

This table shows all possible outcomes:

	1	2	3	4
1	2	3	4	5
2	3	4	5	6
3	4	5	6	7
4	5	6	7	8

The probability distribution table shows the probability for each possible outcome:

X	2	3	4	5	6	7	8
p(x)	$\frac{1}{16}$	$\frac{2}{16}$	$\frac{3}{16}$	$\frac{4}{16}$	$\frac{3}{16}$	$\frac{2}{16}$	$\frac{1}{16}$

This probability distribution can be illustrated in a graph:

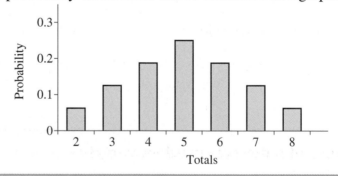

REVISION NOTE

Sometimes a function or rule is given to calculate probabilities.

REVISION NOTE

A table showing values of outcomes and probabilities is called a **probability distribution**.

REVISION NOTE

Not all distributions are symmetrical like this one.

Probability density functions

A discrete random variable may be defined by a function called a probability density function.

EXAMPLE

X is the discrete random variable such that $x = 1, 2, 3, 4$. The probabilities are given by the rule

$$P(X = x) = \frac{x}{10}$$

Write out the probability distribution.

The probability distribution is:

X	1	2	3	4
p(x)	0.1	0.2	0.3	0.4

This can be shown on a graph:

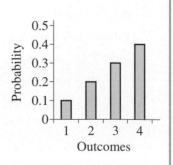

METHOD NOTE

In this case, a calculation is needed to find the probability for each value of x.

$$P(X = x) = \frac{x}{10}$$

So $P(X = 3) = \frac{3}{10} = 0.3$

The cumulative distribution function

This is formed by summing the probabilities up to each value of X. The table below is the cumulative distribution table for the sum obtained when two 4-sided dice are thrown.

X	2	3	4	5	6	7	8
$F(x)$	$\frac{1}{16}$	$\frac{3}{16}$	$\frac{6}{16}$	$\frac{10}{16}$	$\frac{13}{16}$	$\frac{15}{16}$	1

The capital F denotes a cumulative function.

From this table we can see that $P(X \leqslant 5) = \frac{10}{16}$.

We can also see that $P(X = 5) = P(X \leqslant 5) - P(X \leqslant 4) = \frac{4}{16} = \frac{1}{4}$

REVISION NOTE

The cumulative distribution function is a progressive total of probabilities, displayed in a table.

$P(X \leqslant 8) = 1$ because the $p(x)$'s from all outcomes have been summed.

Expectation of a discrete random variable

> The expectation of a discrete random variable can be compared to the mean of a set of data. It is given by
>
> $$E(X) = \Sigma x_i p_i$$

REVISION NOTE

$E(X)$ does not have to be an integer.

$E(X)$ does not have to be one of the values of X.

EXAMPLE

X is the random variable defined as 'the number obtained when a 4-sided die is rolled'.

Calculate the expectation for this random variable.

The table shows the probability distribution.

x	1	2	3	4
$P(X = x)$	$\frac{1}{4}$	$\frac{1}{4}$	$\frac{1}{4}$	$\frac{1}{4}$

To calculate the expectation of X

$$E(X) = \sum_{i=1}^{4} x_i p_i \quad E(X) = (1 \times \tfrac{1}{4}) + (2 \times \tfrac{1}{4}) + (3 \times \tfrac{1}{4}) + (4 \times \tfrac{1}{4})$$
$$= \tfrac{10}{4} = 2.5$$

Variance of a discrete random variable

> The variance of a discrete random variable is given by
> $$Var(X) = E(X^2) - [E(X)]^2$$
> $$= \Sigma x_i^2 p_i - [\Sigma x_i p_i]^2$$
> $Var(X) > 0$ and, as before, standard deviation st. dev. $(X) = \sqrt{Var(X)}$

EXAMPLE

The random variable X has the probability distribution given in the table.

X	2	5
$p(x)$	0.4	0.6

Determine the variance and standard deviation of X.

$E(X) = (2 \times 0.4) + (5 \times 0.6) = 3.8$

$E(X^2) = (2^2 \times 0.4) + (5^2 \times 0.6) = 16.6$

$E(X^2) - [E(X)]^2 = 16.6 - 3.8^2 = 2.16$

$Var(X) = 2.16$ and st. dev $(X) = 1.47$ (2 d.p.)

Discrete random variables

1 Write out the probability distributions for these discrete random variables:

(a) X is the random variable which represents the positive difference between the scores shown when two 6-sided dice are rolled.

(b) Two discs have opposite sides numbered 1 and 2. Both discs are tossed and the random variable Y represents the sum of the numbers shown.

(c) A bag contains 3 black and 4 blue marbles. Two marbles are chosen at random, with replacement. X is the random variable representing the number of blue marbles selected.

2 A discrete random variable Y can take the values

1, 3, 5, 7 and $P(Y = y) = ky$.

Find the value of the constant k.

3 The probability distribution for the random variable X is shown in the table below.

X	2	3	4	5	6
$P(X = x)$	0.02	0.28	0.3	0.18	0.22

Construct the cumulative distribution table for this random variable.

4 For a discrete random variable X, the cumulative distribution table is given below.

X	12	13	14	15
$F(X)$	0.05	0.3	0.75	1

Find:

(a) $P(X \leqslant 13)$ (b) $P(X = 13)$ (c) $P(X > 13)$

5 Calculate the expectation and variance for the random variable X given in question 3.

6 A discrete random variable X may take only two values, each of which has a unique probability.

HINT

Use simultaneous equations

X	2	5
$P(X = x)$	p	q

Given that $E(X) = 3$, find $p = P(X = 2)$ and $q = P(X = 5)$.

Hence find the variance of X.

7 A discrete random variable Y has probability density function

HINT

Draw up a table.

$$P(Y = y) = k(4 - y) \text{ for } y = 0, 1, 2, 3, 4$$

Find the value of the constant k.

6 Discrete probability distributions

A discrete probability distribution can only have whole number outcomes. Each distribution here has its own probability density function and each is used under specific conditions.

The binomial distribution

This is for situations with exactly two outcomes, for example, tossing a coin (heads or tails) or manufacturing a product which is either perfect or faulty.

Conditions of use:

1 There are exactly two mutually exclusive outcomes called *success* and *failure*.

2 Each trial is repeated a fixed number of times.

3 Trials are independent of each other.

REVISION NOTE

You may need to quote the conditions of use in an exam.

The formula for calculating probabilities is

$$P(X = r) = {}^nC_r p^r q^{n-r}$$

where p = the probability of success

q = the probability of failure = $1 - p$

r = the number of successes in n trials

n = the number of trials.

REVISION NOTE

A trial occurs when, for example, one product is made or the coin is tossed once.

For example, 6 coin tosses or 10 articles produced.

We write $X \sim B(n, p)$ to represent the binomial distribution with number of trials n and probability p.

REVISION NOTE

This gives you all the information you need: number of trials and probability of success.

EXAMPLE

Determine the probability of getting 2 heads from 3 tosses of a bent coin which has $P(H) = 0.2$

A success is the coin landing heads, so $p = 0.2$.

There are three tosses so $n = 3$.

We are interested in only getting two heads so $r = 2$.
$q = 1 - p = 0.8$.

Substituting these into the formula gives

$$P(X = 2) = {}^3C_2 \times 0.2^2 \times 0.8^1 = 0.096$$

METHOD NOTE

Look for values of p, n and r in the question, then substitute.

Expected value for $X \sim B(n, p)$

$$E(X) = np$$

Variance for $X \sim B(n, p)$

$$\text{Var}(X) = npq$$

REVISION NOTE

The expected value is the average number of successes in n trials.

The binomial distribution (cont.)

Using cumulative tables

Cumulative tables are used for calculating probabilities over a range of outcomes, for example $P(X < 5)$.

> **EXAMPLE**
>
> For a particular binomial distribution, $p = 0.3$ and there are 7 trials. Possible outcomes (number of successes) are:
>
> | r | 0 | 1 | 2 | 3 | 4 | 5 | 6 | 7 |
> |---|---|---|---|---|---|---|---|---|
> | $p(r)$ | p_0 | p_1 | p_2 | p_3 | p_4 | p_5 | p_6 | p_7 |
>
> Find the probability that (a) $r = 2$ (b) $r \leqslant 2$ and (c) $r \geqslant 2$.
>
> (a) $P(r = 2) = {}^7C_2 \times 0.3^2 \times 0.7^5 = 0.3177$
> (b) $P(r \leqslant 2) = p_0 + p_1 + p_2 = 0.6471$
> (c) $P(r \geqslant 2) = p_2 + p_3 + p_4 + p_5 + p_6 + p_7 = 1 - P(r \leqslant 1)$
> $= 1 - (p_0 + p_1) = 0.6706$

METHOD NOTE

p_0, p_1, p_2, etc. refer to $P(r = 0)$, $P(r = 1)$, etc.

EXAM NOTE

Cumulative binomial tables will be in the handbook or formula book from your exam board.

The calculations for parts (b) and (c) above could be done using **cumulative binomial probability tables**.

Values listed in tables are of the form $P(X \leqslant r)$. Part (b) of the example above can be read from cumulative tables where $n = 7$, $p = 0.3$ and $r = 2$.

METHOD NOTE

Find the table with $n = 7$, go along to $p = 0.3$ and down to $r = 2$. The probability shown here is $P(r \leqslant 2)$. (These tables use x instead of r.)

n	x \ p	0.050	0.100	0.150	1/6	0.200	0.250	0.300	1/3	0.350	0.??
7	0	0.6983	0.4783	0.3206	0.2791	0.2097	0.1335	0.0824	0.0585	?.????	?.???
	1	0.9556	0.8503	0.7166	0.6698	0.5767	0.4449	0.3294	0.2634	0.2338	0.158
	2	0.9962	0.9743	0.9262	0.9042	0.8520	0.7564	0.6471	0.5706	0.5323	0.419
	3	0.9998	0.9973	0.9879	0.9824	0.9667	0.9294	0.8740	0.8267	0.8002	0.710

Expected frequencies

REVISION NOTE

Expected frequencies are the average number of r successes in n trials.

This is the average number of times you would expect a particular outcome to occur over the given number of trials.

> **EXAMPLE**
>
> A bent coin has a probability of 0.65 of landing on heads.
>
> Calculate the expected frequencies for the coin tossed 5 times, by first calculating the probability for each outcome.
>
> | Number of heads, X | $P(X)$ | Expected frequency |
> |---|---|---|
> | 0 | 0.0053 | 0.026 |
> | 1 | 0.0488 | 0.244 |
> | 2 | 0.1811 | 0.906 |
> | 3 | 0.3364 | 1.682 |
> | 4 | 0.3124 | 1.562 |
> | 5 | 0.1160 | 0.580 |
>
> Expected frequency $= np$
> When $x = 1$
> $np = 5 \times 0.0488$
> $= 0.244$
>
> The sum of the expected frequencies should be n.

METHOD NOTE

Probabilities were found using the binomial distribution formula with $n = 5$ and $p = 0.65$.

REVISION NOTE

Expected frequencies are used to check how well a probability model fits experimental data.

Hypothesis testing on a binomial distribution

A hypothesis test is done to establish the validity of the accepted probability of a particular event or experiment.

Every hypothesis test must follow the same pattern:

1 Establish the **null hypothesis** H_0. This takes the form
$H_0 : p =$ (the probability of success for the given experiment)

2 Establish the **alternative hypothesis**, H_1. This is the probability that a success appears to have, from evidence collected or observed. The alternative hypothesis may take one of three forms:

$H_1 : p >$ given probability of success

$H_1 : p <$ given probability of success

$H_1 : p \neq$ given probability of success

The first two forms are one-tailed tests but the third form is a two-tailed test.

3 Establish the sample size, n and the outcome of interest, r.

4 Use the significance level and cumulative tables to establish unacceptable outcomes. These r values form the **critical region**.
The inequality sign in H_1 indicates which tail of the distribution will hold the critical region:

If $H_1 : p <$ probability of success

the critical region will be the lower tail values $0, 1, \ldots$

If $H_1 : p >$ probability of success

the critical region will be the upper tail values $n, n-1, \ldots$

If $H_1 : p \neq$ probability of success

the critical region will contain values from both extremes.

The combined probabilities of the values in the critical region must be less than or equal to the significance level.

EXAM NOTE

Both H_0 and H_1 will be evident from the wording in the question.

See page 100.

EXAM NOTE

Sample size, outcome of interest and significance level will be given in the question.

See page 100.

EXAMPLE

One-tailed test

Spotty brand sweets come in three colours. The maker claims that all colours are equally likely to occur in any packet, but experience has shown that many packets have more blue sweets than any other colour. A packet containing 9 sweets is found to have 5 that are blue. Test at the 10% significance level if this supports the equally likely claim.

$H_0 : p = \frac{1}{3}$ (probability of blue)

$H_1 : p > \frac{1}{3}$

$n = 9$, 10% significance level, $r = 5$

$P(r \geq 6) = 1 - P(r \leq 5)$

$\qquad = 1 - 0.9576 = 0.0424$

$P(r \geq 5) = 1 - P(r \leq 4)$

$\qquad = 1 - 0.8552 = 0.1448$

14.48% > 10% but 4.24% < 10%

Therefore the critical region is $r \geq 6$.

r	$P(r)$
0	0.0260
1	0.1431
2	0.3772
3	0.6503
4	0.8552
5	0.9576
6	0.9917
7	0.9890
8	0.9999
9	1.0000

$r = 5$ is not in the critical region, therefore it is not a rare event. In this case we accept H_0. There is no evidence to suggest that the probability of a blue sweet is greater than $\frac{1}{3}$.

EXAM NOTE

The wording indicates a one-tailed test. Look for words like 'more than', 'less than', 'over-estimate', etc.

METHOD NOTE

These are the cumulative probabilities for a binomial distribution with $p = \frac{1}{3}$ and $n = 9$. The shaded values of r form the critical region.

To find this region, sum the probabilities from the upper extreme down until you reach exactly 10% or just less.

Hypothesis testing on a binomial distribution (cont.)

If the observer of an event or experiment claims that the actual probability of success is not what it is supposed to be, a hypothesis test can be done to test this claim. When the observer cannot tell whether the probability of success has increased or decreased, a two-tailed test is needed.

In this case, extreme values of r at both ends of the distribution will form the critical region.

The significance level must be split into two equal halves and the critical region established.

EXAMPLE

Two-tailed test

A class was given a multiple choice test consisting of 20 questions each with possible answers A, B, C and D. The students told the teacher that they had guessed the answers.

Which test scores would indicate at the 10% significance level that the students had not been guessing?

$H_0 : p = 0.25$, $n = 20$

$H_1 : p \neq 0.25$

10% significance level

In this case the 10% will be split equally between the upper and lower tails.

The two shaded regions shown on the cumulative table are the **critical regions**. If a student scores any of the shaded values of r they did not guess.

The cumulative probability in the lower tail $P(r \leqslant 1) = 0.0243$.

$r = 2$ cannot be included because this would take the combined probability over 5%.

The cumulative probability in the upper tail $P(r \geqslant 9) = 1 - P(r \leqslant 8) = 0.0409$, which is less than 5%.

From this hypothesis test the teacher can see that a score of 0 or 1 indicates that a student consistently chose incorrect answers and was therefore not guessing. A score of 9 or more shows that a student chose correct answers more often than by mere chance.

r	P(r)
0	0.0032
1	0.0243
2	0.0913
3	0.2252
4	0.4148
5	0.6172
6	0.7858
7	0.8982
8	0.9591
9	0.9861
10	0.9961
11	0.9991
12	0.9998
13	1.0000
14	
15	
16	
17	
18	
19	
20	

The Poisson distribution

This distribution is used in situations where events occur at random over a given interval. This interval may be a time period, a distance, or an area for example.

The events must be independent of each other.

The events must have a mean or average rate of occurrence within the given interval. The mean is usually denoted by λ.

REVISION NOTE

The mean or rate of occurrence is the only parameter needed.

$$P_r = P(X = r) = \frac{\lambda^r e^{-\lambda}}{r!} \qquad r = 0, 1, 2, 3, \ldots$$

where λ = mean number of events per given interval.
λ is always a positive number.

REVISION NOTE

This gives the probability that r events occur in the given interval.

We write $X \sim \text{Po}(\lambda)$ to represent the Poisson distribution with mean $= \lambda$.

Common examples of Poisson distributions

1 Phone calls are received at an exchange at a rate of 10 per 30 seconds.
2 The mean number of cars passing my house in a randomly chosen 5-minute period is 35.
3 On average, 12 daisies grow in a randomly chosen square metre of my front lawn.

EXAM NOTE

The main clue here is in the wording. Look for 'rate' or 'mean rate' or 'average'.

EXAMPLE

Water from a pond is being analysed for a particular type of bacteria. It is known that the mean number of bacteria per millilitre is 5. A 1 ml sample is taken from the pond. Find the probability that there will be:

(a) no bacteria   (b) at most 3 bacteria

(a) $P(r = 0) = e^{-5} = 0.006\,74$

(b) $P(r \leqslant 3) = P(r = 0) + P(r = 1) + P(r = 2) + P(r = 3)$

$$= e^{-5} + \frac{e^{-5} \times 5}{1!} + \frac{e^{-5} \times 5^2}{2!} + \frac{e^{-5} \times 5^3}{3!} = 0.2650$$

EXAM NOTE

Simply substitute λ and r into the formula. Check whether you need to know the formula or if it is given in the formula book.

Using cumulative tables

Cumulative Poisson probability tables take the form $P(X \leqslant r)$ where the values are cumulative. We could have used these tables for part (b) of the example above.
Find $\lambda = 5$ along the top of the table and go down the column to $r = 3$. This gives $P(r \leqslant 3)$.

EXAM NOTE

Cumulative Poisson tables will be in the handbook or formula book from your exam board.

NOTE

These tables use x instead of r.

x \ λ	5.00	5.10	5.20	5.30	5.40	5.50
0	0.0067	0.0061	0.0055	0.0050	0.0045	0.0041
1	0.0404	0.0372	0.0342	0.0314	0.0289	0.0266
2	0.1247	0.1165	0.1088	0.1016	0.0948	0.0884
3	0.2650	0.2513	0.2381	0.2254	0.2133	0.2017
4	0.4405	0.4231	0.4061	0.3895	0.3733	0.3575

The Poisson distribution (cont.)

Expected value for $X \sim Po(\lambda)$

$$E(X) = \lambda$$

Variance for $X \sim Po(\lambda)$

$$Var(X) = \lambda$$

EXAM NOTE

You must be careful to check the lengths of all intervals in any questions in case you have to change the parameter.

REVISION NOTE

Practise doing these calculations on your calculator and do not do any rounding until the end.

Changing the interval

In the example on page 101 we looked at bacteria in a 1 ml sample of water. If we want to find the probability of exactly r bacteria in a different volume of water we have to change the value of λ accordingly.

For example, to find the probability of 6 bacteria in 2 ml of water the new parameter would be $2 \times \lambda = 2 \times 5 = 10$.

$\lambda = 5$ per 1 ml. The new parameter $\mu = 10$ per 2 ml.

$$P(r = 6) = \frac{e^{-10} \times 10^6}{6!} = 0.063$$

Sums of independent Poisson distributions

If a question involves two separate situations that must be combined, you can safely assume that these situations are independent. Your new parameter will be a combination of both individual parameters. Take care to ensure that the interval for both is identical.

REVISION NOTE

You can only add parameters when the intervals are equal.

REVISION NOTE

If $X \sim Po(\lambda)$ and $Y \sim Po(\mu)$ then

$$Z = X + Y \sim Po(\lambda + \mu)$$

EXAM NOTE

Always write down what you are going to calculate, to be sure of getting your method marks.

EXAMPLE

Cars pass through a village at the rate of 300 per hour. Lorries pass through the village at the rate of 1 every 5 minutes. What is the probability that exactly 6 vehicles pass through the village in any one-minute period?

The probability that exactly 6 vehicles pass through the village in a particular one-minute period is found by combining the two distributions.

Let X be the Poisson distribution for the cars,
$\lambda = 300$ cars per hour or 5 cars per minute.

Let Y be the Poisson distribution for the lorries,
$\mu = 1$ lorry per 5 minutes = 0.2 lorries per minute.

The combined distribution for cars and lorries is
$Z = X + Y$
with parameter $\mu + \lambda = 5.2$ per minute.

Therefore

$$P(Z = 6) = \frac{5.2^6 e^{-5.2}}{6!} = 0.151 \ (3 \text{ d.p.})$$

Discrete uniform distribution

Here every outcome x_i, of a random variable X, is equally likely.

There are k possible outcomes.

REVISION NOTE

All probabilities are equal. Examples are tossing a fair coin, rolling a fair die.

Each outcome has the probability:

$$p(x_i) = \frac{1}{k}$$

EXAMPLE

X is the discrete random variable defined as 'the score obtained when a fair die is rolled'.
Write out the probability distribution of X.

In this case $k = 6$.

Value of x	1	2	3	4	5	6
Probability	$\frac{1}{6}$	$\frac{1}{6}$	$\frac{1}{6}$	$\frac{1}{6}$	$\frac{1}{6}$	$\frac{1}{6}$

The graph of the distribution shows why it is called 'uniform'.

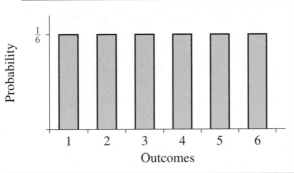

NOTE

All the columns are the same height because all the outcomes are equally likely.

Expected value $E(X) = \Sigma x p(x)$

Variance $Var(X) = \Sigma x^2 p(x) - [E(X)]^2$

Geometric distribution

This arises in situations where the experiment stops at the first success, e.g. playing board games where you have to throw a six to start.

Probability of success is p.
Probability of failure is q.

The distribution is given by $P(X = x) = pq^{x-1}$ for $x \geq 1$.

Expected value $E(X) = \dfrac{1}{p}$

Variance $Var(X) = \dfrac{q}{p^2}$

EXAMPLE

In Ludo you need to throw a six to start.
Find the probability of starting on the third turn.

$p = \frac{1}{6}$, $q = \frac{5}{6}$, so $P(X = 3) = \frac{1}{6} \times \left(\frac{5}{6}\right)^2 = \frac{25}{316}$

METHOD NOTE

There are two failures before the first success, i.e. two throws which are not a six before the first six.

Discrete probability distributions

1 A random variable X takes the values 1, 2, 3, 4, 5. It has a discrete uniform probability distribution.

   (a) Draw up a table for the probability distribution.

   (b) Calculate $P(X \leqslant 3)$.

   (c) Calculate the expectation and the variance for this distribution.

2 One in ten chocolate bars contains a token for a free bar of chocolate. A shopper buys chocolate bars until she finds a winning token.
   Find:

   (a) The probability that a token is found in the 1st bar.

   (b) The probability that a token is found in the 4th bar.

   (c) The probability that the shopper will have to buy at least four bars before finding a token.

3 A packet of 12 tulip bulbs is said to contain 30% yellow tulips and the rest red.

   (a) Find the probability that upon flowering exactly 2 tulips are yellow.

   (b) Find the probability that upon flowering fewer than 3 tulips are yellow.

   (c) Find the expected number of yellow tulips.

4 A customer services phone line receives an average of 30 calls per hour. Calculate the probability that:

   (a) In any 5-minute period there are no calls.

   (b) In a 10-minute period there are more than 7 calls.

5 A garden centre states that packets of mixed crocus bulbs consist of 40% yellow flowers and 60% purple flowers.

   (a) Calculate the probability that exactly 6 have purple flowers from a packet of 15 bulbs.

   (b) Calculate the probability that 6 or fewer have purple flowers.

   (c) An amateur gardener suspects that the probability of a purple bulb is less than 60%. He finds that only 7 flowers in a packet of 15 are purple. Perform a hypothesis test at the 10% significance level and state your conclusions.

   (d) Would your conclusion have been the same if exactly 6 flowers were purple?

6 In a practice session before a competition, a marksman hits the bullseye with probability 0.4.

   In the competition the marksman shoots at 20 targets and hits a bullseye 11 times.

   Perform a hypothesis test at the 10% significance level to ascertain whether or not the probability that he hits the bullseye has changed.

7 Expectation algebra

Definitions

Expectation

$E(X) = \Sigma xp$ or $\Sigma xP(X = x)$

REVISION NOTE

$E(X)$ does not have to be an integer.

In all cases, the numerical value sought is the 'long-term average value'.

See examples in Discrete random variables, page 95.

Variance

$\text{Var}(X) = E(X^2) - [E(X)]^2 = \Sigma x^2 p - (\Sigma xp)^2$

$\text{Var}(X) \geqslant 0$ and Standard deviation $= \sqrt{\text{Var}(X)}$

Expectations of functions of random variables

$E[g(X)] = \Sigma g(x)p \quad \text{Var}[g(X)] = \Sigma [g(x)]^2 p - \left[E[g(x)]\right]^2$

These rules are an extension of the original formulae for expectation and variance (see page 95). They enable us to calculate expectation and variance for cases where the given random variable is transformed into a second random variable with the same probabilities.

REVISION NOTE

Forgetting to subtract $(E[g(x)])^2$ is a very common mistake.

EXAMPLE

X is a discrete random variable which can take the values shown in the table below.

x	$0°$	$45°$	$90°$
$P(X = x)$	$\frac{1}{3}$	$\frac{1}{3}$	$\frac{1}{3}$

Find $E(\sin X)$ (the expected value of $\sin X$) for this probability distribution.

$E(\sin X) = (\sin 0 \times \frac{1}{3}) + (\sin 45° \times \frac{1}{3}) + (\sin 90° \times \frac{1}{3})$

$\qquad = 0.569$ (3 d.p.)

NOTE

In the example the random variable X is an angle, but we want the expectation and variance of the new variable: $\sin x$.

METHOD NOTE

The rules are the same. Just use $g(X)$ instead of X.

More formulae I

$\qquad E(X + a) = E(X) + a \qquad \text{Var}(X + a) = \text{Var}(X)$

Similarly,

$\qquad E(X - a) = E(X) - a \qquad \text{Var}(X - a) = \text{Var}(X)$

REVISION NOTE

Adding or subtracting a constant a to each X value does not change the spread of the distribution.

EXAMPLE

The table below shows the probability distribution for a random variable X. Given that $E(X) = 2.9$ and $\text{Var}(X) = 2.89$, find $E(X + 3)$ and $\text{Var}(X + 3)$.

x	0	1	3	5
$P(X = x)$	0.1	0.2	0.4	0.3

Using the rules:

$E(X + 3) = (3 \times 0.1) + (4 \times 0.2) + (6 \times 0.4) + (8 \times 0.3) = 5.9$

$\text{Var}(X + 3) = (9 \times 0.1) + (16 \times 0.2) + (36 \times 0.4)$

$\qquad\qquad\qquad + (64 \times 0.3) - 5.9^2 = 2.89$

METHOD NOTE

Adding 3 to each x gives the new distribution

x	3	4	6	8
$P(x)$	0.1	0.2	0.4	0.3

REVISION NOTE

You should check that using the formulae with the transformed distribution gives the same result as using the rules.

METHOD NOTE

This shows that $E(3X) = 3E(X)$

$3E(X) = 3 \times 2.9$
$\qquad = 8.7$

and

$Var(3X) = 3^2 Var(X) = 9 \times 2.89$
$\qquad\qquad = 26.01$

EXAM NOTE

Forgetting to subtract the square of the expectation is a very common mistake.

More formulae II

$$E(aX) = aE(X) \qquad\qquad Var(aX) = a^2\,Var(X)$$

EXAMPLE

X is a discrete random variable with the probability distribution shown in the table below. Find $E(3X)$ and $Var(3X)$, given that $E(X) = 2.9$ and $Var(X) = 2.89$.

x	0	1	3	5
$P(X = x)$	0.1	0.2	0.4	0.3

$E(3X) = (0 \times 0.1) + (3 \times 0.2) + (9 \times 0.4) + (15 \times 0.3)$
$\qquad\quad = 8.7$
$Var(3X) = (0 \times 0.1) + (3^2 \times 0.2) + (9^2 \times 0.4)$
$\qquad\qquad\quad + (15^2 \times 0.3) - 8.7^2$
$\qquad\quad = 26.01$

When the previous two rules are combined we find that:

$$E(aX + b) = aE(X) + b \text{ and } Var(aX + b) = a^2Var(X)$$

It can also be shown that:

$$E[g(X) + h(X)] = E[g(X)] + E[h(X)]$$

EXAMPLE

In a fairground game a player pays 20p to toss 3 coins. The player wins 10p for each heads he obtains.
Calculate the mean and variance of the showman's net gain.

The probability distribution is shown in this table. X is the number of heads thrown.

X	0	1	2	3
$P(X)$	0.125	0.375	0.375	0.125

X is the random variable for the number of heads obtained, so the net loss will be given by

$\qquad Y = 20 - 10X$

$E(X) \quad = (0 \times 0.125) + (1 \times 0.375) + (3 \times 0.375)$
$\qquad\qquad\quad + (1 \times 0.125)$
$\qquad\quad = 1.5$

$E(X^2) = (0 \times 0.125) + (1 \times 0.375) + (4 \times 0.375)$
$\qquad\qquad\quad + (9 \times 0.125)$
$\qquad\quad = 3$

$Var(X) = 3 - 1.5^2 = 0.75$

$E(Y) \quad = 20 - 10E(X) = 20 - (10 \times 1.5)$
$\qquad\quad = 5 \qquad$ (average gain of 5p per player)

$Var(Y) = (-10)^2Var(X) = 100 \times 0.75$
$\qquad\qquad = 75$

METHOD NOTE

$20 - 10X$ is how much the showman will end up with after the player has a turn.

It is much quicker to use the formulae than to calculate $E(20 - 10X) = \Sigma(20 - 10x)p$

Expectation algebra

1 Calculate $E(X)$ and $Var(X)$ for the random variable below.

x	1	3	5	7	9	11
$P(X = x)$	$\frac{1}{12}$	$\frac{1}{6}$	$\frac{1}{4}$	$\frac{1}{3}$	$\frac{1}{12}$	$\frac{1}{12}$

From your results find:

(a) $2E(X)$ (b) $Var(2X)$

(c) $E(3X)$ (d) $E(2X - 1)$

(e) $Var\left(\dfrac{X}{4} + 2\right)$

2 Given that $E(Y) = 3$ and $Var(Y) = 2$, find the expectation and variance of:

(a) $Y - 1$ (b) $3Y$

(c) $5 - Y$ (d) $2Y + 3$

3 Given that $E(X) = -2$ and $Var(X) = 4$ find:

(a) $E(2X)$ (b) $Var(2X)$

(c) $E(-3X)$ (d) $Var(-3X)$

Show also that $E(2X) + E(-3X) = E(-X)$

4 A fairground game is played as follows. Two fair dice are rolled and the scores are summed. If the player scores 12, she receives £5. She receives £1 if she scores 2 or 3. For any other score she pays 50p to the showman.

> **HINT**
> Write the distribution for the loss in a table.

What is the expected win for the player?

5 A car salesman receives commission of £45 for each car he sells as well as a fixed daily income of £35. The number of cars he sells in a day varies from 0 to 4.

The random variable X represents the number of cars sold in a day and the table shows the probability distribution for X.

x	0	1	2	3	4
$P(X = x)$	0.4	0.3	0.2	0.06	0.04

The salesman's daily income Y is given by

$$Y = 45X + 35$$

> **HINT**
> Find $E(X)$. Use this to find $E(Y)$, then the expected value for the week.

(a) Calculate his expected daily income.

(b) Assuming that he works five days each week, calculate his expected weekly income.

(c) Calculate the variance of his daily income.

8 Continuous random variables

Definition

A discrete random variable may only take a given number of values, but a continuous random variable may take any value in a given interval.

A continuous random variable will be described by a function called the **probability density function (pdf)**. This has the properties:

1 $f(x) \geq 0$

2 $P(a < X < b) = \int_a^b f(x)\,dx$

3 $\int_c^d f(x)\,dx = 1$

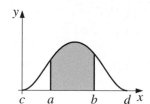

REVISION NOTE

Probabilities will always be calculated over an interval for continuous random variables.

REVISION NOTE

The area under the curve must equal 1 for f(x) to be a continuous random variable.

METHOD NOTE

This random variable is only defined for values of x between 1 and 3.

EXAM NOTE

It is a good idea to sketch the graph of the function if you can. A diagram often makes the problem easier to understand.

The shaded area will give the probability required.

EXAMPLE

The continuous random variable X has pdf

$$f(x) = \begin{cases} \frac{3}{26} x^2 & 1 < x < 3 \\ 0 & \text{otherwise} \end{cases}$$

Determine $P(X < 2)$.

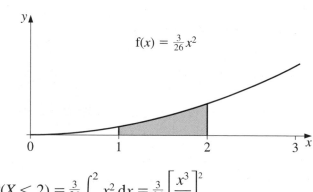

$$P(X < 2) = \tfrac{3}{26} \int_1^2 x^2\,dx = \tfrac{3}{26}\left[\frac{x^3}{3} \right]_1^2$$

$$= \tfrac{3}{26}\left[\tfrac{8}{3} - \tfrac{1}{3} \right]$$

$$= \tfrac{3}{26} \times \tfrac{7}{3} = \tfrac{7}{26}$$

REVISION NOTE

Normal distributions arise in any situation where you have most of your measurements clustered around a central value. For example, heights and weights of animals of any species, lengths of leaves on some trees, size of pebbles on the beach, etc.

The normal distribution

We write $N(\mu, \sigma^2)$ to represent the normal distribution with a mean of μ and a variance of σ^2.

A normal distribution is a uni-modal, symmetrical, continuous distribution defined by its parameters μ (the mean) and σ^2 (the variance). These symbols represent the mean and variance for the population.

The standard normal distribution

All normal distributions can be related to a single reference distribution called the **standard normal distribution**.

This has mean = 0 and variance = 1 and is denoted by $Z \sim N(0, 1)$.

The probability density function of Z is denoted by $\phi(z)$ where

$$\phi(z) = \frac{1}{\sqrt{2\pi}}\, e^{-\frac{1}{2}z^2}$$

$$-\infty < z < \infty$$

This function looks like this.

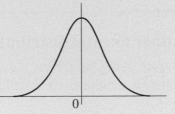

REVISION NOTE

This function cannot be integrated so we use tables to calculate all probabilities. Note the symbol Φ used for the probabilities.

$$P(Z \leq a) = \int_{-\infty}^{a} \frac{1}{\sqrt{2\pi}}\, e^{-\frac{1}{2}z^2}$$

for any value $z = a$.

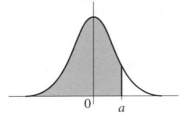

EXAM NOTE

Always draw a diagram like this to show the area you are trying to find.

$\Phi(a) = P(Z \leq a)$ is the shaded area under the curve.

EXAMPLE

Find $P(Z < 1.5)$ given $Z \sim N(0, 1)$.

On normal distribution tables, go to $z = 1.5$ on the $\Phi(z)$ table.

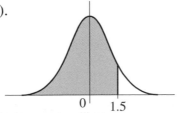

EXAM NOTE

Because this distribution is continuous, $P(Z < a) = P(Z \leq a)$ and $P(Z \geq a) = P(Z > a)$

z	.00	.01	.02	.03	.04	.05	.06	.07	.08	.09	1	2	3	4	5
1.5	9332	9345	9357	9370	9382	9394	9406	9418	9429	9441	1	2	4	5	6
1.6	.9452	9463	9474	9484	9495	9505	9515	9525	9535	9545	1	2	3	4	5
1.7	.9554	9564	9573	9582	9591	9599	9608	9616	9625	9633	1	2	3	3	4

$P(Z < 1.5) = 0.9332$

EXAM NOTE

Normal distribution tables will be in the handbook or formula book from your exam board.

Finding $\Phi(-z)$

Tables give values for $\Phi(z)$ for non-negative values of z. $\Phi(-z)$ is found using the symmetry of the graph of the distribution.

$\Phi(-z) = 1 - \Phi(z)$

because the total area under the curve = 1.

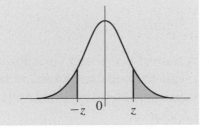

REVISION NOTE

The shaded areas are equal. The table will only give the value for the area to the left of z.

The standard normal distribution (cont.)

EXAMPLE

Find $P(Z < -1.5)$ given $Z \sim N(0, 1)$.

$\Phi(-1.5) = 1 - \Phi(1.5)$ (Diagram shows that these areas are equal.)

$P(Z < -1.5) = 1 - 0.9332 = 0.0668$

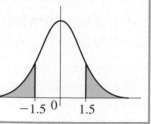

$-1.5 \quad 0 \quad 1.5$

Other normal distributions

Before using the tables, values from other normal distributions must be standardised as follows:

If $X \sim N(\mu, \sigma^2)$, then

$$P(X < x) = P\left(Z \leqslant \frac{x - \mu}{\sigma}\right) = \Phi\left(\frac{x - \mu}{\sigma}\right)$$

EXAMPLE

Given that $X \sim N(3, 25)$, find $P(X < 4.2)$.

$$z = \frac{x - \mu}{\sigma} \Rightarrow z = \frac{4.2 - 3}{5} = 0.24$$

Using the tables:

$\Phi(0.24) = 0.5948$
$P(X < 4.2) = 0.5948$

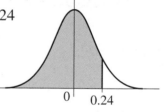

$0 \quad 0.24$

Finding μ or σ from given information

EXAMPLE

The random variable X has a normal distribution $N(\mu, \sigma^2)$ where $\mu = 6.4$ and $P(X < 7) = 0.7580$.
Find the value of σ.

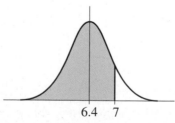

$6.4 \quad 7$

$P(X < 7)$ is equivalent to the shaded area.
$\Phi(z) = 0.7580$
so $z = 0.7$

$$z = \frac{7 - 6.4}{\sigma} = 0.7 \qquad \text{so} \qquad 7 - 6.4 = 0.7\sigma$$

Simplifying gives

$\sigma = 0.857$

This method can be used to find μ if you are given a distribution where σ is known but μ is not.

Continuous random variables

1 Verify that each of the following is a pdf and sketch its graph:

(a) $f(x) = 1$, $(0 < x < 1)$

(b) $f(x) = 2x$, $(0 < x < 1)$

(c) $f(x) = |x|$, $(|x| < 1)$

2 Given that

$$f(x) = \begin{cases} k(4 - x) & 1 \leqslant x \leqslant 3 \\ \\ 0 & \text{otherwise} \end{cases}$$

Find the value of:

(a) k (b) $P(X < 2)$ (c) $P(1.5 < X < 2.5)$

3 The pdf of a random variable X is:

$f(x) = kx^2 (1 - x)$, $(0 < x < 1)$

(a) Determine k.

(b) Find the probability that X is less than $\frac{2}{3}$.

> **HINT**
>
> Sketch the graph. For (a), the area under the graph must equal 1.

4 Given that

$$f(x) = \begin{cases} 4 + x & -k < x < k \\ \\ 0 & \text{otherwise} \end{cases}$$

find the value of k for which this is a pdf.

5 If $X \sim N(30, 20)$ find:

(a) $P(X < 33)$ (b) $P(X > 28)$ (c) $P(28 < X < 33)$

6 If $X \sim N(84, 12)$ find:

(a) $P(79 < X < 87)$ (b) $P(X > 87)$ (c) $P(X < 79 \text{ or } X > 87)$

7 Let $X \sim N(27, 16)$.

(a) If $P(X < a) = 0.75$, find a.   (b) If $P(X < b) = 0.25$, find b.

(c) If $P(X > c) = 0.3$, find c.   (d) If $P(X > d) = 0.89$, find d.

8 If $X \sim N(43, \sigma^2)$ and $P(X < 62) = 0.873$, find σ.

9 If $X \sim N(\mu, 25)$ and $P(X < 31.4) = 0.504$, find μ.

9 Sampling

A sample is a subset of the population.

Why consider a sample?

1 Cost-effective
2 Time saving.
3 Accuracy – errors can be more effectively controlled in a small project than in a large one.
4 More detail – more effort can be used on each member of the sample.
5 Destructive testing – e.g. a company manufacturing a particular component for the aircraft industry is not going to test all its components to destruction.

Before you choose your sample you may have a list of all members of the population. This list is called a **sampling frame**.

Random sampling methods

Here each member of the population is equally likely to be chosen in the sample. No section of the population is excluded at any time during the process. This method does not guarantee a sample which is representative of the population.

Method

Members of the population are assigned a unique number. A computer is used to generate random numbers. The members with those numbers are chosen as part of the sample. (This is like drawing names out of a hat.)

Non-random sampling methods

Systematic sampling

This does not necessarily give a representative sample.

Method

A 10% sample of a population is required. An ordered list of the entire population is obtained. A random starting number between 1 and 10 is generated. The member of the population with this number will be the starting point. Every 10th member of the population after this starting point is sampled.

The only random part of this process is choosing the starting point. Every member has the same chance of being included in the sample until the starting point is chosen. This immediately excludes large sections of the population.

Cluster sampling

A small population is identified as being typical of all other similar populations. (e.g. a penguin rookery could represent all penguins). All members of this cluster are sampled. This is particularly useful for a widely spread population.

REVISION NOTE

A census includes every member of the population but it is expensive and time consuming.

REVISION NOTE

For random sampling and systematic sampling, the sampling frame would be an ordered list of the entire population.

For cluster sampling (see below) the sampling frame may be a list of all suitable clusters, e.g. all penguin rookeries.

METHOD NOTE

Your calculator will also produce random numbers.

REVISION NOTE

This method could be used when analysing sentence lengths in a children's book. Choose a random starting page and sentence number on that page.

REVISION NOTE

One street in a housing estate could represent all other streets, one pod of whales could represent all other pods of whales, etc. This is used a lot with wildlife studies.

Non-random sampling methods (cont.)

Stratified sampling

Here the sub-samples are taken from distinct sections of the population. (e.g. in a school, each year level would be sampled separately). Each section of the population should be listed so that each member of that section is equally likely to be chosen. The final sample has the same proportions of each section as there are in the population.

Method

This method is best shown through an example.

EXAMPLE

Suppose a college has 620 sixth form students enrolled on A-Level courses. A 5% representative sample of students is to be chosen to complete a questionnaire about college facilities. 220 students are studying Science A-Levels, 100 are studying Arts A-Levels and the rest are studying Humanities. Explain how you would collect the sample and describe the composition of the sample.

5% of 620   = 31 students.

5% of each subject area must be sampled.

5% of 220   = 11

5% of 100   = 5

5% of 300   = 15

total sample = 31

Using random sampling, 11 students are chosen from the list of Science students, 5 from the Arts and 15 from the Humanities.

The sample has the same proportions of each group as the population.

EXAM NOTE
A question on stratified samples will require calculations of percentages.

METHOD NOTE
By taking 5% of each stratum (subject group) the sample is representative of the population.

REVISION NOTE
You need to know how to calculate the numbers needed for a sample of this kind.

Quota sampling

This method is used for practical reasons. It is like strata sampling but is quite subjective. The selection method is left to the sampler, which could introduce some bias.

Method

The sampler is instructed to obtain a sample of, say, 50 women, to include a given number of women in each age group. A sampler could stand in the High Street and ask passers-by if they would like to take part in a survey. They ask the age of the volunteer and if the sampler does not have enough women of this age group the volunteer would be included in the sample. If the sampler already had her quota for this age group, the volunteer would not be surveyed.

REVISION NOTE
This method is often used in telephone surveys where the caller will ask your age category before beginning the survey.

10 Correlation and regression

Pearson's product moment correlation coefficient

Pearson's product moment correlation coefficient is denoted by r where

$$r = \frac{S_{xy}}{S_x S_y}$$

S_x = standard deviation of x

S_y = standard deviation of y

$$S_{xy} = \frac{1}{n}\Sigma xy - \bar{x}\,\bar{y}$$

EXAM NOTE

A scatter diagram will give you a good idea of whether the correlation is positive or negative.

REVISION NOTE

$r = -1$ means perfect negative correlation.

$r = 1$ means perfect positive correlation.

The correlation coefficient r, is a measure of the **strength of the relationship** between the two variables x and y.

$-1 \leq r \leq 1$. A positive value for r implies that as x increases y increases. A negative value for r implies that as x increases y decreases. When r is close to zero there is little or no correlation. When r is close to 1 or -1 there is strong positive or negative correlation respectively.

EXAMPLE

Calculate Pearson's product moment correlation coefficient for the following data set.

x	35	15	10	25	30	20	50	5	40	25
y	30	30	25	40	50	10	55	20	40	50

Calculating r:

$\Sigma x = 255$ $\Sigma x^2 = 8255$

$\Sigma y = 350$ $\Sigma y^2 = 14\,150$

$\Sigma xy = 10\,150$

METHOD NOTE

Calculate summary statistics like these using regression mode on your calculator.

METHOD NOTE

You need to use the formula for S_{xy}, because there is no button for it on your calculator.

$$S_{xy} = \frac{1}{n}\Sigma xy - \bar{x}\,\bar{y}$$

$$= \tfrac{1}{10} \times 10\,150 - 25.5 \times 35 = 122.5$$

$$S_x = \sqrt{\frac{1}{n}\Sigma x^2 - \bar{x}^2}, \qquad S_y = \sqrt{\frac{1}{n}\Sigma y^2 - \bar{y}^2}$$

$$S_x = 13.12 \quad \text{and} \quad S_y = 13.78$$

Moderate positive correlation.

$$r = \frac{122.5}{13.12 \times 13.78} = 0.677$$

The scatter graph for this data also shows a positive correlation.

If $r = 1$ or -1 the scatter graph would show a straight line with all points on the line.

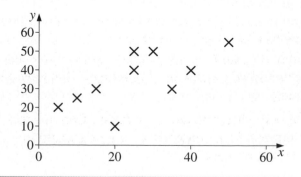

Linear regression

When a scatter diagram shows a very strong correlation between two variables, we may wish to find the straight line that approximates the relationship between the two variables. This straight line will have the equation $y = ax + b$ and is called the **y on x regression line**.

You can find this as follows:

$$y - \bar{y} = \frac{S_{xy}}{S_x^2}(x - \bar{x})$$

$$S_{xy} = \frac{\Sigma xy}{n} - \bar{x}\,\bar{y} \quad \text{and} \quad S_x^2 = \frac{\Sigma x^2}{n} - \bar{x}^2$$

METHOD NOTE

Once you have put your data into your calculator using regression mode, various combinations of buttons will give you $\Sigma x^2, \Sigma xy$, Σx, etc.

REVISION NOTE

Every regression line passes through $(\bar{x}, \bar{y})$.

EXAMPLE

Calculate the linear regression line of y on x for the table of values shown.

x	22	27	32	37	42	47
y	75	67	62	55	45	39

Program the data into your calculator. The summary statistics are:

$\Sigma xy = 11\,201 \qquad \Sigma x = 207 \qquad \Sigma x^2 = 7579$

$\Sigma y = 343 \qquad n = 6$

$S_{xy} = \dfrac{11\,201}{6} - \dfrac{207}{6} \times \dfrac{343}{6} = -105.42$

$S_x^2 = \dfrac{7579}{6} - \left(\dfrac{207}{6}\right)^2 = 72.92$

$\bar{x} = 34.5 \qquad \bar{y} = 57.2$

$y - 57.2 = \dfrac{-105.42}{72.92}(x - 34.5)$

$y = -1.45x + 107$

This shows a negative relationship between the variables.

EXAM NOTE

When doing exam questions it is better to use summary statistics from your calculator than to read the values of a and b straight from the calculator.

EXAM NOTE

Get your regression line into the form $y = ax + b$, by moving all terms except y to the right hand side.

Once you have found the regression line equation, values of y may be estimated for given values of x.

For the example above, you could estimate the value of y when $x = 40$ but not the value of y for $x = 50$. This is because the recorded values of x range from 22 to 47 and 50 is outside this range. As you do not know what happens outside the given range for x, you cannot use the regression line to make estimates.

Similarly, you could not find the y-value corresponding to $x = 20$ because this is outside the valid region.

A regression line may be used:

1 When one variable depends upon another (e.g. a child's height depends upon age, so calculate the height on age regression line).

2 When both variables are random and from normal distributions (e.g. comparison of maths exam mark with physics exam mark).

REVISION NOTE

For **1** you may calculate both y given an x-value and x given a y-value from the y on x regression line.

For **2** you can only calculate y given x from the y on x regression line.

Correlation and regression

1   The weekly wages, £x, of employees in a certain company and their weekly expenditure £y for sports, cinema, travel, entertainment, etc. are as follows:

x	360	370	490	500	385	420	525	480	485	540	450	550	800
y	50	80	150	65	130	90	140	95	60	160	110	150	240

Calculate Pearson's product moment correlation coefficient for the data. Describe the type of correlation, if any, that exists between wages and expenditure.

2   A farmer wants to establish the relationship between the yield in tonnes per hectare of his crop and the price per tonne that he is paid for his crop. He collects the following information over a period of 6 years.

Year	Yield tonnes per hectare, x	Price £ per tonne, y
1	0.8	100
2	0.5	110
3	2.4	150
4	2.1	130
5	1.8	140
6	2.6	165

Calculate the linear regression equation for the data and predict the price per tonne when the yield is 1.5 tonnes per hectare.

3   The manager of a local shop is looking into the relationship between the maximum temperature on some days in June and the number of cans of soft drinks sold on those days. He has collected the following data during seven randomly selected days of observations.

Temperature, $x°$ C	25	28	29	31	34	36	37
Number of cans sold, y	162	196	203	252	300	311	329

(a) Calculate the correlation coefficient between the temperature and the number of cans sold.

(b) Find the linear regression line for y on x.

(c) Use your equation from (b) to predict the sales on a day when the temperature is 27° C.

(d) Can you make a prediction for sales when the temperature is 20° C?

4   In an experiment, a student recorded the temperature of a liquid in a flask.

t (minutes)	0	1	2	3	4	5	6	7	8
$\theta°$ C	65	55	50	42	38	35	34	33	32

She decided to calculate the regression line for temperature on time.

With reference to a scatter diagram, explain why this would not be valid.

Answers and hints to solutions – Statistics

Exercise 1: Discrete data

1 (a) m = 31, UQ = 33, LQ = 25, IQR = 8 (b) m = 10, UQ = 15, LQ = 6, IQR = 9

(c)
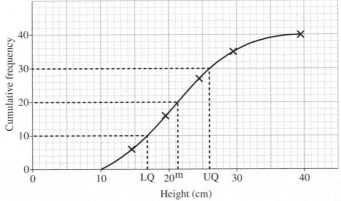

100% of (b) temperatures were below the lower extreme of (a) temperatures.
IQR almost equal; highest value for (b) is less than lowest value for (a).

2 Class A: n = 22, Σx = 1609, Σx^2 = 122 467, $\bar{x}$ = 73.14, s = 14.76
Class B: n = 20, Σx = 1180, Σx^2 = 73 634, $\bar{x}$ = 59, s = 14.17

Class A had higher marks than Class B, but the spread was about the same for both classes.

3 (a) $\bar{x}$ = 1.6, s = 1.575
(b) Combined: $\Sigma x = 40 + 1.2 \times 25$, $\Sigma x^2 = 126 + 25(1.6^2 + 1.2^2)$.
$\bar{x}$ = 1.4, s = 1.6

4

2	0	2	3	4	5	⑥	8			m = 40.5
3	1	8	9							LQ = 26
4	0	m 1	3	5	6	6	⑥	7	7	UQ = 46
5	0	3	5							IQR = 20

n = 22   5|0 = 50

Exercise 2: Continuous data

1 (a) (b)

Class interval	Class width	Frequency density
0–9.5	9.5	1.579
9.5–19.5	10	3.5
19.5–29.5	10	3.7
29.5–49.5	20	0.4
49.5–89.5	40	0.125

(c) $\bar{x}$ = 21.488   s = 14.272

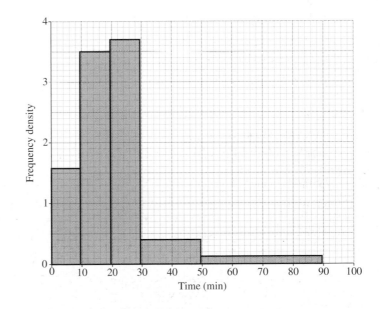

2 (a) 10–14: 6 plants; 25–29: 8 plants
(b)

	Cumulative frequency
<14.5	6
<19.5	16
<24.5	27
<29.5	35
<39.5	40

(c) m ≈ 21.5
UQ ≈ 26
LQ ≈ 16.75
IQR ≈ 9.25

3 $\bar{x} = 72$ min, $s = 4.2$ min

4 $y = \dfrac{(x-1)}{10}$ so $\bar{x} = 10\bar{y} + 1$ and $s_x = 10s_y$

y	f	fy	y^2	fy^2
1	3	3	1	3
2	5	10	4	20
3	11	33	9	99
4	7	28	16	112
5	4	20	25	100

$\bar{y} = 3.1\dot{3}$, $s_y = 1.147$ $\bar{x} = 32.\dot{3}$, $s_x = 11.47$

Exercise 3: Orderings

1 $\dfrac{{}^{15}C_2 \times {}^{12}C_1}{{}^{27}C_3}$ or $3 \times \left(\frac{15}{27} \times \frac{14}{26} \times \frac{12}{25}\right) = \frac{28}{65} = 0.4308$

2 456 976 000

3 (a) $7! = 5040$ (b) $\dfrac{6!}{2!} = 360$ (c) $\dfrac{7!}{2!3!} = 420$

4 ${}^{10}C_3 = 120$

5 (a) $\frac{5}{42} = 0.119$ (b) $\frac{1}{21} = 0.048$ (c) 0

6 (a) ${}^{9}C_4 = 126$ (b) ${}^{7}C_2 = 21$ (c) ${}^{7}C_3 + {}^{7}C_3 = 70$ (d) ${}^{7}C_4 = 35$

7 (a) $\frac{5}{126}$ (b) $\frac{1}{126}$ (c) $\frac{1}{126}$

Exercise 4: Probability

1 (a) $\frac{3}{35}$ (b) Yes since $P(A) \times P(B) = \frac{3}{35}$ (c) $P(A') = \frac{4}{7}$, $P(B') = \frac{4}{5}$

2 (a) 0   (b) No since $P(C) \times P(D) \neq 0$ (c) Yes since $P(C \cap D) = 0$

3 $\dfrac{P(\text{odd and mult. of 3})}{P(\text{odd})} = \dfrac{\frac{2}{12}}{\frac{6}{12}} = \frac{2}{6} = \frac{1}{3}$

4 $P(A \cap B) = \frac{5}{72}$, $P(B|A) = \frac{5}{24}$
Not independent because $P(A|B) \neq P(A)$

5 (a) $\frac{1}{50}$ (b) $\frac{9}{50}$ (c) $\frac{27}{95}$ (d) $\frac{19}{29}$

6 (a) 0.2   (b) 0.08   (c) 0.72

Exercise 5: Discrete random variables

1 (a)

x	0	1	2	3	4	5
p	$\frac{6}{36}$	$\frac{10}{36}$	$\frac{8}{36}$	$\frac{6}{36}$	$\frac{4}{36}$	$\frac{2}{36}$

(b)

x	2	3	4
p	$\frac{1}{4}$	$\frac{1}{2}$	$\frac{1}{4}$

$\Sigma p = 1$

(c)

x	0	1	2
p	$\frac{9}{49}$	$\frac{24}{49}$	$\frac{16}{49}$

2 $k = \frac{1}{16}$

3

x	2	3	4	5	6
$F(x)$	0.02	0.30	0.60	0.78	1

4 (a) 0.30   (b) 0.25   (c) 0.70

5 $E(X) = 4.3$, $Var(X) = 1.33$

6 $\left.\begin{array}{l} p + q = 1 \\ 2p + 5q = 3 \end{array}\right\}$ $p = \frac{2}{3}$, $q = \frac{1}{3}$

7 $k = \frac{1}{10}$

Exercise 6: Discrete probability distributions

1 (a)

x	1	2	3	4	5
p	0.2	0.2	0.2	0.2	0.2

(b) $P(X \leq 3) = 0.6$ (c) $E(X) = 3$, $Var(X) = 2$

2 (a) 0.1   (b) 0.0729   (c) 0.729   (d) $E(X) = 10$ Geometric

3 (a) 0.1678   (b) 0.2528   (c) 3.6   Binomial

4 (a) 2.5 calls per 5 minutes $P(0) = e^{-2.5} = 0.082$

     (b) 5 calls per 10 minutes $P(r > 7) = 1 - P(r \leq 7) = 0.1334$ Poisson

5 (a) 0.0612   (b) 0.0950

     (c) $H_0: p = 0.6$ $n = 15$ 10% sign. level   Binomial
        $H_1: p < 0.6$ $r = 7$
        $P(r \leq 7) = 0.2131 = 21.3\%$ (not a rare event) $21.3\% > 10\%$
        Accept H_0. There is no evidence to suggest that P(purple) is less than 0.6.

     (d) $P(r \leq 6) = 0.095 = 9.5\%$ (more rare than we are happy to accept)
        Reject H_0. There is evidence to suggest that P(purple) is less than 0.6. Therefore the conclusion is different to (c).

6 $H_0: p = 0.4$
   $H_1: p \neq 0.4$
   $n = 20$, $r = 11$, 10% significance level
   From tables, critical values are 0, 1, 2, 3, or $r = 13, 14, 15, \ldots, 20$
   11 is not in either critical region, so accept H_0: p has not changed.

Exercise 7: Expectation algebra

1 $E(X) = 5.8\dot{3}$, $Var(X) = 6.97\dot{2}$
     (a) $11.6\dot{6}$ (b) $27.\dot{8}$ (c) $17.4\dot{9}$ (d) $10.\dot{6}$ (e) 0.436

2 (a) 2, 2   (b) 9, 18   (c) 2, 2   (d) 9, 8

3 (a) -4 (b) 16   (c) 6   (d) 36

4

X	£5 (12)	£1 (2)	£1 (3)	$-£0.50$
$P(X)$	$\frac{1}{36}$	$\frac{1}{36}$	$\frac{2}{36}$	$\frac{32}{36}$

     $E(X) = -0.1944 = -19$ pence loss per turn.

5 (a) £81.80   (b) £409   (c) 2426.76

Exercise 8: Continuous random variables

1 (a) 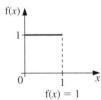   (b) 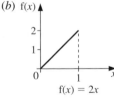   (c)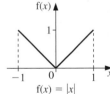

$f(x) = 1$ $f(x) = 2x$ $f(x) = |x|$

2 (a) $k = \frac{1}{4}$ (b) $\frac{5}{8}$ (c) 0.5

3 (a) $k = 12$ (b) $\frac{16}{27}$

4 $k = \frac{1}{8} \int_{-k}^{k} (4 + x)\,dx = 1$ Using tables from the formula book.

5 (a) 0.7489   (b) 0.6725   (c) 0.4214

6 (a) 0.7323   (b) 0.1932   (c) 0.2677

7 (a) 29.698   (b) 24.302   (c) 29.096   (d) 22.092

8 $\sigma = 16.657$ **9** $\mu = 31.36$

Exercise 10: Correlation and regression

1 $\Sigma x^2 = 3\,259\,075$, $\Sigma y^2 = 210\,650$, $\Sigma xy = 800\,050$,
   $\Sigma x = 6355$, $\Sigma y = 1520$, $n = 13$, $r = 0.8046$
   Positive correlation.
   Employees earning higher wages tend to spend more.

2 Price $= 87.94 + 26.21 \times$ yield.   Price $= £127.26$ $y = 26.21x + 87.94$

3 (a) 0.99   (b) $y = 14.54x - 206.54$ (c) 186   (d) No, it is outside the valid range.

4 Scatter diagram shows **non-linear** relationship between time and temperature.

Mechanics Topics:

1 Distance, velocity and acceleration
Vectors and scalars   121
Addition and subtraction of vectors   121
Multiplication by a number   122
Resolving vectors – components   122
Magnitude of a vector   122
Direction of a vector   123
Kinematics   123
Constant acceleration formulae   123
Constant acceleration graphs   125
Constant acceleration and vectors   126
Calculus and kinematics   127
Calculus with vectors   128
Exercise 1: Distance, velocity and
    acceleration   129

2 Forces and Newton's laws
Force diagrams   130
Newton's laws of motion   131
Mass and weight   131
Using Newton's second law   132
Connected particles and Newton's
    third law   133
Using Newton's third law   134
Exercise 2: Forces and Newton's laws   137

3 Equilibrium
Equilibrium and force diagrams   138
General result for three forces   140
A particle in equilibrium on an
    inclined plane   140
Equilibrium problems using vector
    methods   142
Exercise 3: Equilibrium   143

4 Friction
The coefficient of friction, μ 144
Forces applied at an angle   145
Friction on rough inclined planes   146
Motion up an inclined plane   147
Exercise 4: Friction   148

5 Momentum and impulse
Momentum   149
Impulse   149
Collisions   150
Exercise 5: Momentum and impulse   151

6 Projectiles
Horizontal projection   152
Projectiles launched at an angle to
    the horizontal   153
Symmetry of the path of a projectile   154
Projection at an angle from a point
    above the plane   154
Time of flight   155
Range of a projectile   155
Maximum range of a projectile   156
Maximum height of a projectile   156
Equation of the trajectory   157
Exercise 6: Projectiles   158

Answers and hints to solutions   159

1 Distance, velocity and acceleration

Vectors and scalars

Ordinary numbers, measurements of time, and measures of mass are all examples of **scalars**.

> A scalar quantity has a size (known as its magnitude) but no particular direction.

The usual rules of arithmetic – such as addition and subtraction – are applied to scalar quantities.

Velocity, acceleration, force are all quantities which have a definite **direction**, as well as a **magnitude**.

> A vector quantity is one that has both magnitude and direction.

There are special rules for addition and subtraction of vectors. Vectors can also simply be multiplied by a number. Multiplication of a vector by a vector requires a special definition.

Notation

Vectors are normally written in printed texts using bold format: for example **a**, **PQ**, etc.

When vectors are handwritten, they are shown by using an underline: for example a̲, P̲Q̲, etc.

When vectors are indicated by a pair of letters such as **PQ**, the vector is the displacement from P to Q.

A single letter for a vector, such as **a**, sometimes indicates the vector displacement from an origin to a point **A**, so that $\mathbf{a} \equiv \mathbf{OA}$.

Addition and subtraction of vectors

When vectors are added, the outcome of the addition is called a **resultant vector**. The magnitude and direction of the resultant is given by a **triangle rule** of addition.

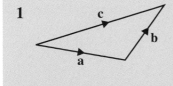

In diagram **1**.

$$\mathbf{c} = \mathbf{a} + \mathbf{b}$$

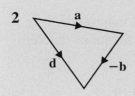

In diagram **2**, the vector **−b** has the same magnitude, but opposite direction to vector **b**. So the diagram represents

$$\mathbf{d} = \mathbf{a} + (-\mathbf{b})$$
$$= \mathbf{a} - \mathbf{b}$$

> **REVISION NOTE**
>
> Arithmetic rules such as multiplication and division are not always appropriate, even with scalars.

> **REVISION NOTE**
>
> Multiplication of a vector by a vector is not in the AS-Level specification.

> **NOTE**
>
> Other notations are also used, such as $\overrightarrow{PQ}$.

Multiplication by a number

Multiplication of a vector by a number (a scalar) changes the length of the vector, but its direction remains parallel to the original vector. If the number is negative, the direction of the vector is reversed.

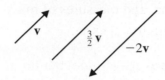

The diagram shows a vector **v**, and the vector which arises when it is multiplied by scalars $\frac{3}{2}$ and -2.

All three vectors are parallel.

METHOD NOTE

In 3 dimensions, vectors can be resolved into 3 components all at right angles to each other.

Resolving vectors – components

Vectors are often **resolved** into **components** – two other vectors perpendicular to each other that can be added to give the original vector as a **resultant**.

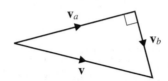

In the diagram, the two component vectors $\mathbf{v}_a$ and $\mathbf{v}_b$ are at right angles, and add together to give the resultant vector **v**.

REVISION NOTE

This idea can also be applied in 3 dimensions. Then the unit vector parallel to the z-axis is **k**.

In many cases, and especially in mechanics problems, it is convenient to resolve vectors into components in the directions of the x and y axes. Vectors in the directions of the x and y axes are so commonly used that they have a special notation: they are written as multiples of **i** and **j**.

i and **j** are **unit vectors** – vectors of length 1, parallel to the axes.

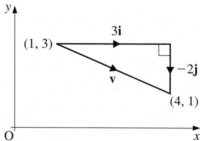

In the diagram, the vector **v** is resolved into two component vectors, 3**i** and -2**j**, parallel to the axes. So the vector **v** can be written in two forms:

$$\mathbf{v} = 3\mathbf{i} - 2\mathbf{j} = \begin{pmatrix} 3 \\ -2 \end{pmatrix}$$

3**i** $-$ **j** is written in component form.

$\begin{pmatrix} 3 \\ -2 \end{pmatrix}$ is written in column vector form.

EXAM NOTE

Column vector form, and component form are completely interchangeable. You can use whichever you prefer, but you should try to give an exam answer in the form the question uses.

Magnitude of a vector

The size of a vector – its length – is called its magnitude. It is calculated from its component form using Pythagoras' theorem.

For example, for the vector $3\mathbf{i} - 2\mathbf{j}$

$$\text{magnitude } |\mathbf{v}| = \sqrt{3^2 + (-2)^2} = \sqrt{13} \approx 3.61$$

Note that the magnitude is a number (a scalar).

Direction of a vector

Finding the direction of a vector involves applying simple trigonometry.

EXAMPLE

Find the direction of the vector $\mathbf{v} = 3\mathbf{i} - 2\mathbf{j}$

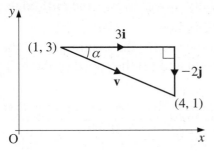

From the diagram

$$\tan \alpha = \frac{-2}{3}$$

$$\approx -0.667$$

$$\Rightarrow \quad \approx -33.7°$$

So vector $\mathbf{v}$ has direction $33.7°$ below the x-axis.

REVISION NOTE

This answer could also be written as 'direction of $326.3°$ to the x-axis' using the convention of measuring angles anti-clockwise from the positive x-axis.

Kinematics

Kinematics studies the relationships between position, velocity and acceleration. Many exam questions use vector notation, and examples of these are covered on page 128.

One special case is when the motion of the particle involved all takes place in one straight line, under the effect of a **constant acceleration**. Then the problems can be simplified without using vectors, and there is a special set of formulae to use.

Constant acceleration formulae

When a particle moves in a straight line under the effect of a constant acceleration, the following formulae can be used:

$v = u + at$ where t = the time elapsed
$s = ut + \frac{1}{2}at^2$ u = the initial velocity
$v^2 = u^2 + 2as$ v = the final velocity
                                     a = the constant acceleration
                                     s = the displacement

REVISION NOTE

u, v, a, s are really vector quantities. They can be positive or negative. The time t can also have negative values, meaning 'times before the instant chosen as the starting time when $t = 0$'.

If displacements are in metres and times in seconds, then velocities are m s^{-1}, accelerations are in m s^{-2}.

EXAMPLE

A car starts to move from rest in a straight line, and reaches a speed of 30 m s^{-1} after 15 seconds.

Find (a) how far it has travelled, and
(b) how fast it is travelling after it has travelled 100 m.

$$v = u + at \Rightarrow 30 = 0 + a(15)$$
$$\Rightarrow a = 2 \,(\text{m s}^{-2})$$

(a) $\quad s = ut \times \frac{1}{2}at^2 = 0 \times 2 + \frac{1}{2} \times 2 \times (15)^2 = 225$ m

(b) $\quad v^2 = u^2 + 2as \Rightarrow v^2 = (0)^2 + 2 \times 2 \times 100 = 400$
$$\Rightarrow v = 20 \text{ m s}^{-1}$$

Make sure you write the correct units.

METHOD NOTE

Assume that the acceleration remains constant and use the constant acceleration formulae.

Constant acceleration formulae (cont.)

Interpreting negative values is an important aspect of questions involving the constant acceleration formulae.

A man stands at the edge of a cliff, 60 metres above the sea below, and throws a ball vertically upwards over the cliff edge with an initial velocity of $5\,\text{m s}^{-1}$.

Find (a) when the ball lands in the sea (b) the maximum height of the ball above the starting point and (c) the speed of the ball as it hits the water.

We must first define some variables. Here, we will assume that the acceleration due to gravity is $10\,\text{m s}^{-2}$ vertically downwards. We also fix the starting point where $s = 0$ as the point from which the ball is thrown. Take vertically upwards as positive, so that the acceleration throughout is $a = -10$. The initial velocity is given by $u = +5$.

(a) When the ball lands in the sea, we have $s = -60$, and so

$$s = ut + \tfrac{1}{2}at^2$$
$$\Rightarrow \quad -60 = 5t + \tfrac{1}{2}(-10)t^2$$
$$\Rightarrow \quad 5t^2 - 5t - 60 = 0$$
$$\Rightarrow \quad t^2 - t - 12 = 0$$
$$\Rightarrow \quad (t - 4)(t + 3) = 0$$

so the ball hits the sea after 4 seconds.

(b) The highest point is when the ball stops moving upwards and begins to move downwards again, i.e. the velocity is instantaneously zero.

$$v = u + at$$
$$\text{so } v = 0 \Rightarrow \quad 0 = 5 + (-10)t$$
$$\Rightarrow \quad 10t = 5$$
$$\Rightarrow \quad t = \tfrac{1}{2}$$

So the highest point is reached $\tfrac{1}{2}$ second after the ball is thrown. The height is

$$s = ut + \tfrac{1}{2}at^2$$
$$= 5\left(\tfrac{1}{2}\right) + \tfrac{1}{2}(-10)\left(\tfrac{1}{2}\right)^2$$
$$= \tfrac{5}{4}$$

So the maximum height is 1.25 m above the start.

(c) The speed at which the ball hits the sea could be found using either u, a and t with $v = u + at$

or u, a and s with $v^2 = u^2 + 2as$

Using the first formula,

$$v = 5 + (-10)(4)$$
$$= -35$$

and so the ball hits the sea moving vertically downwards at a speed of $35\,\text{m s}^{-1}$.

Constant acceleration graphs

The relationship between displacement, velocity and acceleration means that some constant acceleration questions are best tackled by sketching a graph.

The relationship is summarised in this table:

Gradient	Which graph?	Area beneath
velocity ⟵ ——— displacement–time		
acceleration ⟵ ——— velocity–time ———⟶		displacement
acceleration–time ———⟶		velocity

REVISION NOTE

This suggests a **calculus** approach. We look at that on page 127.

EXAMPLE

An underground train accelerates from a station to reach a speed of 20 m s^{-1} after 10 seconds. It travels at constant speed for 20 seconds, and then takes another 20 seconds to stop at the next station.

(a) How far apart are the two stations?
(b) What is the deceleration of the train when it is stopping?

The diagram shows a velocity–time graph for the whole motion.

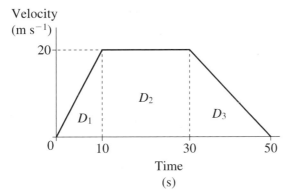

(a) The areas marked D_1, D_2 and D_3 are the distances travelled in the three parts of the motion, and can be found using simple area formulae.

$$D_1 = \tfrac{1}{2} \times 10 \times 20 = 100 \text{ (m)}$$
$$D_2 = 20 \times 20 = 400 \text{ (m)}$$
$$D_3 = \tfrac{1}{2} \times 20 \times 20 = 200 \text{ (m)}$$

So the stations are $100 + 400 + 200 = 700$ m apart.

(b) The deceleration is given by the gradient of the last part of the graph, so deceleration $= \dfrac{(-20)}{20} = -1 \text{m s}^{-2}$.

METHOD NOTE

In drawing a graph with straight line sections, we are assuming constant acceleration.

EXAM NOTE

Don't look for difficult ways of doing the calculation. Just use area formulae for triangles, rectangles, and trapezia where they fit.

NOTE

A 'deceleration' is just a negative acceleration.

In exam questions on constant acceleration, the calculations are usually easy because you will always use the familiar formulae. But there are two important steps before you start the question.

1 Check to make sure that constant acceleration is a valid assumption.
2 For each part of the motion, write down clearly which variables in the equation you know (and which you need to find).

Calculus and kinematics (cont.)

METHOD NOTE

Speed v is a maximum when

$\dfrac{dv}{dt} = 0$ so solve, then substitute

into the expression for v.

(b) The train reaches maximum speed when $a = 0$.

$$v = 5 + \frac{1}{2}t^2 - \frac{1}{400}t^2 \Rightarrow a = \frac{dv}{dt} = \frac{1}{2} - \frac{1}{200}t$$

So when $a = 0$, $\dfrac{1}{2} = \dfrac{1}{200}t \Rightarrow t = 100$

So maximum speed is when $t = 100$,

$$v = 5 + \tfrac{1}{2} \times 100 - \frac{1}{400} \times 100^2 = 30 \text{ m s}^{-1}$$

METHOD NOTE

There are two solutions for
$v = 0$, but one of them is
negative, and is not appropriate
for the model. So we choose the
positive value…

(c) The train stops when $v = 0$, so we must solve a quadratic equation

$$v = 0 \Rightarrow 200 + 200t - t^2 = 0$$

$$\Rightarrow t = \frac{-200 \pm \sqrt{(200)^2 - 4(-1)(2000)}}{2(-1)}$$

$$= 100 \pm \sqrt{12\,000} = 209.5 \text{ (1 d.p.)}$$

…and the value 209.5 becomes
the top limit for this integral –
when the train stops. The bottom
limit, zero, is the time when the
train left the tunnel.

So the distance travelled from the tunnel to the second signal where the train stops is given by

$$s = \int_0^{209.5} 5 + \tfrac{1}{2}t - \frac{1}{400}t^2 \, dt$$

$$= \left[5t + \tfrac{1}{4}t^2 - \frac{1}{1200}t^3 \right]_0^{209.5}$$

$$= 4360 \text{ m} \quad (\text{to 3 s.f.})$$

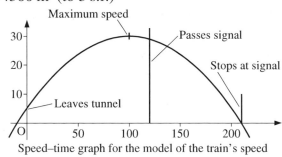

Speed–time graph for the model of the train's speed

Calculus with vectors

The next example shows how calculus can be used with vector questions in kinematics.

METHOD NOTE

Because **i** and **j** are constant
vectors, all you do here is
differentiate the functions of t.

Since the coefficient of **i** in the
velocity is constant, when you
differentiate there is no **i**
component.

EXAMPLE

The position of a particle with respect to the origin at time t is given by $\mathbf{r} = 3t\mathbf{i} + t^3\mathbf{j}$. Find the velocity and acceleration of the particle after one second, giving the magnitude and angle to the x-axis of each.

$$\mathbf{r} = 3t\mathbf{i} + t^3\mathbf{j}$$

$$\Rightarrow \mathbf{v} = \frac{d\mathbf{r}}{dt} = 3\mathbf{i} + 3t^2\mathbf{j}$$

$$\Rightarrow \mathbf{a} = \frac{d\mathbf{v}}{dt} = \frac{d^2\mathbf{r}}{dt^2} = 6t\mathbf{j}$$

So $t = 1 \Rightarrow \mathbf{v} = 3\mathbf{i} + 3\mathbf{j}$ and $\mathbf{a} = 6\mathbf{j}$

So the speed of the particle is $\sqrt{18} = 4.24 \text{ m s}^{-1}$ at $45°$ to the positive x-axis, and the acceleration is 6 m s^{-2} at $90°$ to the positive x-axis.

REVISION NOTE

Integration using vectors works
exactly the same way. But as in
all kinematics, you must be
careful about the limits of the
integration.

Distance, velocity and acceleration

(Use $g = 9.8 \text{ m s}^{-2}$ where appropriate.)

1   A car starts from rest and accelerates at a constant rate of 1.5 m s^{-2}.

Find how fast it is travelling and how far it has moved after 12 seconds.

How fast is it travelling when it has moved 50 metres from the starting point?

2   A man throws a stone vertically upwards from a beach at 20 m s^{-1}, and it passes the top of the cliff after 1.5 seconds.

How high is the cliff?

The stone lands on the edge of the cliff as it falls again. When does it land and how fast is it then moving?

3   A bus leaves a stop, and accelerates at a constant rate of 1 m s^{-2} until it reaches 15 m s^{-1}. It then travels for some time at constant speed, and then decelerates at a constant 0.5 m s^{-2} to stop at the next stop, exactly 1 kilometre from the first.

Sketch the velocity–time graph and find the time spent travelling at constant speed.

4   A particle starts from $(3, 2)$ with initial velocity equal to $3\mathbf{i} - \mathbf{j}$ and moves with constant acceleration given by $-\mathbf{i} + \mathbf{j}$.

Find the speed and position of the particle after 6 seconds.

5   A particle moves with speed

$$\mathbf{v}(t) = t + t^2$$

along the x-axis. When $t = 0$, the particle is at $x = 2$.

(a) Find an expression for $s(t)$, the position at time t.

(b) Show that when $t = -2$, the particle is at $x = \frac{4}{3}$.

6   A bicycle is being ridden at 10 m s^{-1} in a straight line at constant speed, when it reaches a section of the cycle path covered in sand. This causes the bicycle to decelerate at a constant 1.5 m s^{-2}.

How far does the bicycle travel before it stops?

7   The position of a particle at t seconds is given by the formula:

$$\mathbf{r} = 2t^2\mathbf{i} - (3t - 1)\mathbf{j}$$

(a) Find the position of the particle after 0, 1, 2 and 3 seconds and sketch its path on a graph.

(b) Find formulae for $\mathbf{v}$, the velocity, and $\mathbf{a}$, the acceleration, of the particle.

(c) Show that the acceleration of the particle is constant throughout its motion. Calculate this acceleration.

(d) Find the magnitude and direction of the velocity when $t = 1$.

2 Forces and Newton's laws

Force diagrams

This unit deals with **dynamics** – how forces cause accelerations. Later Mechanics units look at **statics** – how systems of forces occur which keep particles in equilibrium, with zero acceleration.

In most questions on statics or dynamics, a clear diagram will help to ensure that important details are not missed. Often it is best to draw more than one diagram – one showing the 'real' picture to help you understand the model you are using, and then others showing the forces involved, the masses, and the accelerations that result.

For example, these diagrams could help in analysing a question about a lorry of mass m accelerating at a up a hill under the effect of a driving force D, with weight W, normal contact force N, and friction force F. The units are all omitted in this example, but must always be picked to be consistent in a question.

METHOD NOTE

Keep to simple, consistent units: metres, kilograms, seconds and Newtons.

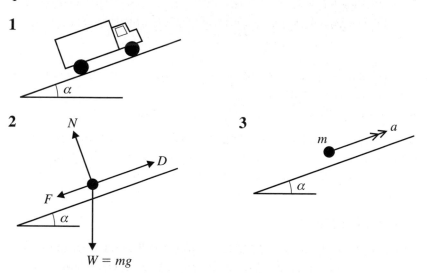

1

2

3

$W = mg$

Use clear conventions for the diagrams: *solid* arrowheads to represent forces, and *double* arrowheads to represent accelerations.

METHOD NOTE

As we see later, this helps make it easy to equate force, and mass–acceleration.

In the example above, separate diagrams are drawn to show:

2   All the forces acting on the lorry, treated as a particle.

3   The mass, and the consequent acceleration.

The diagrams are carefully drawn with the directions clearly shown, so that **components** of the forces can be resolved if necessary.

REVISION NOTE

In this example, there will be an equal but opposite 'contact force' pushing 'downwards' on the road, but that is not relevant to the problem.

Show only the forces that act on the 'particle' you are working with. So in diagram **2** the friction is shown acting opposing the acceleration up the hill, and the normal contact force is shown acting in an 'upwards' direction at right angles to the road.

Newton's laws of motion

In AS-Level, Newton's laws of motion are used to model the behaviour of all sorts of bodies, such as cars, trains, children, snooker balls and anything else that an examiner can think of. In almost every case we make the assumption that we are dealing with **particles**.

METHOD NOTE

This is an example of mathematical modelling. Always be clear about what simplifications you are making.

Newton's laws can be taken as:

1   A particle remains stationary, or continues to move at constant speed in a straight line, unless an external force is applied.

2   The acceleration of the particle is proportional to the magnitude of the resultant external force on the particle, and in the direction of the resultant external force.

   The result of this law is so important that it is usually called the equation of motion, and written as:

   $$F = ma$$

3   If two particles exert contact forces on each other, the forces are equal in magnitude, and opposite in direction.

NOTE

Newton's second law actually says something more complicated than this, but these 'rules' explain how it applies at AS-level.

The third law is applied in situations where the particles exert forces on each other, either:

● by touching directly, such as a car parked on a road, or

● by being connected by a string in tension, or a rod in tension or compression, such as a person lifting a load by pulling on a rope over a pulley, or a car pulling a caravan with a towbar.

REVISION NOTE

You don't need to learn Newton's laws by heart, but you do need to understand how and when to apply them. Most important, you must remember

$$F = ma$$

Mass and weight

Mass is a property of any particular body, and is independent of whether the body is on Earth, on the Moon, or in outer space. **Mass is measured in kilograms, usually written as kg**.

Weight is the force that acts on the particle as a result of gravity. The formula connecting mass and weight for a particle is

$$W = mg$$

where **g is the acceleration due to gravity**. The value of g is usually taken to be 9.8 m s^{-2} at the Earth's surface.

Weight is a force, and **force is measured in Newtons, usually written as N**.

The difference between mass and weight is important in every mechanics question and the two ideas are often confused in everyday life. The convention recommended on page 130 should help. On a forces diagram weight is shown as a vector with magnitude and direction. Mass is shown on the mass–acceleration diagram as a property of the particle, with no direction.

EXAM NOTE

Always read the question carefully. Sometimes you are told to use $g = 10$, which is easier. The true value is nearer 9.81, but because of the shape of the Earth, it is not really constant everywhere. Just occasionally, a question may be set 'on the moon' to test your understanding by using a different value of g.

Using Newton's second law

The equation of motion $F = ma$ is applied in the examples below.

EXAMPLE

A particle of mass 7 kg rests on a smooth table. Two horizontal forces of 9 N and 4 N act on the particle in opposite directions.

Find the acceleration of the particle.

First draw two diagrams – one of the forces involved, and one showing mass and acceleration.

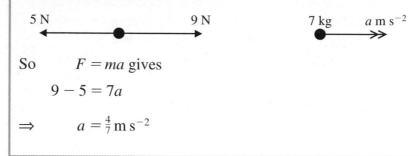

So $\quad F = ma$ gives

$$9 - 5 = 7a$$

$$\Rightarrow \quad a = \tfrac{4}{7}\,\text{m s}^{-2}$$

In this example, the motion took place on a '*smooth horizontal table*', which is a common **mathematical model** in exam questions. This allows you to ignore the vertical forces in the calculation.

EXAMPLE

A load with mass 50 kg is accelerated upwards by a crane at $1.5\,\text{m s}^{-2}$ when it is first lifted vertically from the deck of a ship. What is the tension in the cable connecting the load to the crane?

In this problem, work out the weight of the load, and then draw diagrams of forces, and mass and acceleration.

The weight of the load is

$$W = mg$$
$$= 50 \times 9.8$$
$$= 490\,\text{N}$$

So $\quad F = ma$ gives

$$T - W = ma$$
$$T - 490 = 50 \times 1.5$$
$$T = 490 + 75$$
$$= 565\,\text{N}$$

and so the tension in the cable is 565 Newtons.

Using Newton's second law (cont.)

The next example is another common examiners' model that avoids the need to consider weight. It involves a spacecraft in outer space, where the model ignores any gravity.

A spacecraft in deep space is acted on by a force given by $(8000\mathbf{i} - 3000\mathbf{j})$ N where $\mathbf{i}$ and $\mathbf{j}$ are unit vectors along the axis of the spacecraft, and at right angles to it. If the mass of the spacecraft is 5 tonnes, give the magnitude and direction of its acceleration.

The mass of 5 tonnes is 5000 kg.

So applying Newton's second law in vector form:

$$\mathbf{F} = m\mathbf{a}$$

$$\Rightarrow \quad 8000\mathbf{i} - 3000\mathbf{j} = (5000)\mathbf{a}$$

$$\Rightarrow \qquad\qquad \mathbf{a} = 1.6\mathbf{i} - 0.6\mathbf{j} \text{ m s}^{-2}$$

From the diagram

$$|\mathbf{a}| = \sqrt{(1.6)^2 + (0.6)^2}$$

$$= \sqrt{2.92} = 1.71 \text{ m s}^{-2} \text{ (3 s.f.)}$$

and $\quad \alpha = \tan^{-1}\left(\dfrac{-0.6}{1.6}\right)$

$$= \tan^{-1}(-0.35) = 20.6° \text{ (3 s.f.)}$$

So the acceleration is 1.71 m s^{-2}, at an angle of 20.6° to the positive direction of $\mathbf{i}$ in the negative $\mathbf{j}$ direction.

This question does not involve 'weight'. But $F = ma$ still applies, here written in vector format.

This is a vector equation, where $\mathbf{F}$ and $\mathbf{a}$ are vectors. The mass is *not* a vector, so we can divide the coefficients of $\mathbf{i}$ and $\mathbf{j}$ by 5000 to get a vector result for $\mathbf{a}$.

Connected particles and Newton's third law

Many exam questions on Newton's third law involve bodies 'connected' together, by direct contact or by strings or rods.

In such questions:

- Identify clearly where Newton's third law applies, and so where the 'equal and opposite' forces take effect.
- Draw separate diagrams for the connected particles, showing the forces acting on each and the masses and accelerations involved.
- Analyse the motion using $F = ma$ for the separate parts of the connected system.

In some questions you can also use $F = ma$ on the whole system, but you must be careful to check whether that is valid before you do.

This trick works when you are not asked to find any of the internal forces in the strings or rods.

Using Newton's third law

Look at the next three typical, but different, examples.

A man of mass 70 kg is standing in a lift when it accelerates upwards with an acceleration of 1.5 m s^{-2}. The lift cage has a mass of 500 kg, and the upwards force is provided by a cable fixed to the lift cage.

Find (*a*) the tension in the cable when the lift accelerates, and (*b*) the contact force between the lift and the man during the acceleration.

Because we want first to find the tension in the cable, and *not* an **internal** force, we can **treat the whole system as one**. So the first calculation treats the lift and passenger as a single particle.

For the whole system of lift and man:

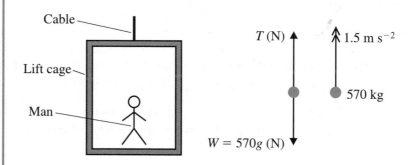

METHOD NOTE

This diagram shows forces, mass and acceleration for the complete system of lift cage plus man, taken together as one particle.

METHOD NOTE

The weight *W* here is the total weight of the lift plus passenger.

NOTE

We have assumed that $g = 9.8 \text{ m s}^{-2}$ throughout these examples.

(*a*) For the whole system,
$$T - W = 570 \times 1.5$$
$$T - 570 \times 9.8 = 570 \times 1.5$$
$$T - 5586 = 855$$
$$T = 6441$$

So during the acceleration, the tension in the supporting cable is approximately 6440 N.

(*b*) Considering the equation of motion for just the passenger:

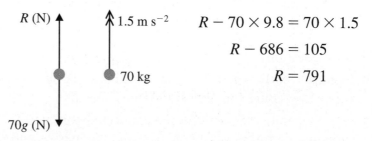

$$R - 70 \times 9.8 = 70 \times 1.5$$
$$R - 686 = 105$$
$$R = 791$$

REVISION NOTE

This diagram and calculation includes only the man and the two forces acting on him – his weight downwards and the contact force upwards. So the diagram includes his mass, 70 kg.

During the acceleration, the **contact force** between the lift cage and the man is approximately 790 N.

So the man will experience an upwards force of 790 N, which is greater than his weight, 686 N. This explains the momentary sensation of feeling 'heavy' that you experience as a lift starts upwards.

Using Newton's third law (cont.)

A railway locomotive is decelerating a single truck along a horizontal railway track at 0.3 m s^{-2}. The truck has a mass of 40 tonnes, and the locomotive has a mass of 70 tonnes.

Assuming that there is a resistance force of 5000 N to the motion of the truck, find (*a*) the braking force provided by the locomotive, and (*b*) the force in the coupling rod.

In this question, all the important forces are horizontal. Although (very large) vertical forces act – the weight of the locomotive and the truck, and the normal reaction upwards from the rails – there is no vertical acceleration throughout the motion so we can assume that these forces cancel each other out. They play no part in the solution.

We also assume that the only significant friction force to consider is the resistance on the truck.

(*a*) For the whole system, there are two horizontal external forces – the resistance force on the truck, tending to slow it down, and the braking force from the locomotive, which is slowing down both the locomotive and the truck.

The diagram below assumes the train is moving to the left, so all the forces and the acceleration act to the right.

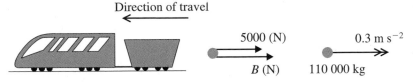

Direction of travel

5000 (N)   0.3 m s^{-2}

B (N)   110 000 kg

So, for the whole system,

$$B + 5000 = 110\,000 \times 0.3$$
$$= 33\,000$$
$$\Rightarrow \qquad B = 28\,000$$

The locomotive provides a braking force of 28 000 N.

(*b*) To find the force in the coupling rod, look at just the truck. If the force on the truck from the coupling rod is *T* N:

T (N)   5000 (N)   0.3 m s^{-2}

40 000 kg

$$5000 - T = 40\,000 \times 0.3$$
$$= 12\,000$$
$$\Rightarrow \qquad T = -7000$$

The minus sign means that the force *T* acts in the opposite direction to that suggested by the diagram. The coupling is in **compression**, with equal and opposite forces of 7000 N acting backwards on the truck, and forwards on the locomotive. So the truck is pushing the locomotive forwards as it tries to slow the train.

METHOD NOTE

With all these **modelling assumptions**, we can now solve the problem in the same way as the last example. First treat the whole system as a single particle, and then apply Newton's laws to just the truck.

NOTE

The numbers are big, but be sure to be consistent. Here we use **1 tonne = 1000 kg** to make sure we use consistent units.

EXAM NOTE

Be careful about directions. Here we take the direction of the **acceleration** as positive. You could use the direction of travel instead – but be consistent.

EXAM NOTE

The direction of *T* is not clear before we start. But the maths will sort it out. If we had chosen the opposite direction for *T* on the diagram, we would have got a positive answer, as long as we were careful to get the sign right in the equation of motion.

Using Newton's third law (cont.)

EXAMPLE

A particle A of mass 2 kg lies at rest on a horizontal table, connected by a light string to another particle B of mass 7 kg. The string passes over a smooth, light pulley at the edge of the table, and particle B hangs vertically below the pulley. If the resistance to sliding of particle A is 10 N, find the initial acceleration of the system, and the tension in the string.

This question is different from the last two examples, because for particle B we must consider the weight.

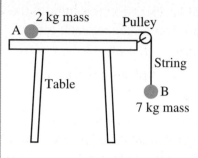

But for particle A, because the motion is horizontal, we can ignore the weight of A, and the normal reaction from the table.

Therefore we draw separate pairs of diagrams for each particle, and write down equations of motion for each.

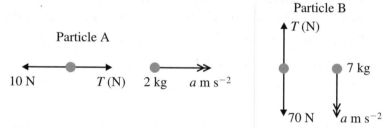

Using the directions of the accelerations as the positive directions in each case,

Particle A gives: $T - 10 = 2a$ (1)

Particle B gives: $70 - T = 7a$ (2)

Solving these two equations simultaneously gives values for T and a. Adding equations (1) and (2),

$$60 = 9a$$

$$\Rightarrow \qquad a = \frac{60}{9} = \frac{20}{3} \, \text{m s}^{-2}$$

and $(1) \Rightarrow T = 2 \times \dfrac{20}{3} + 10 = \dfrac{70}{3} \, \text{N}$

So the acceleration of the two particles is $6.67 \, \text{m s}^{-2}$, with particle A moving to the right, and particle B downwards. The tension in the string is $23.33 \, \text{N}$.

In each of the these examples, the calculation was easy. The important step is to analyse carefully the separate parts of the connected system. Separate and clear diagrams are very important.

Notice how in each case the separate diagrams for forces and for mass–acceleration make it easy to create the equations of motion. The elements in the force diagram appear on the opposite side of the equation to those in the mass–acceleration diagram.

Forces and Newton's laws

(Use $g = 9.8 \text{ m s}^{-2}$ where appropriate.)

1 A particle of mass 10 kg is accelerated vertically upwards at 0.5 m s^{-2}.

What is the upwards force on the particle?

2 A car of mass 750 kg tows a caravan of mass 500 kg along a level road. All the external driving force, or where appropriate, all the external braking force, acts on the car. Each of the car and caravan separately experience a constant resistance to motion of 500 N.

Find the force in the towbar, and the driving or braking force as appropriate, in the following circumstances:

(*a*) The car and caravan accelerate at 1 m s^{-2}.

(*b*) The car and caravan travel at constant speed.

(*c*) The car and caravan decelerate at 1.5 m s^{-2}.

3 A particle A of mass 10 kg lies on a table and is connected by a light inextensible string to another particle B of mass 5 kg. The string passes over a smooth pulley at the edge of the table, and the particle B hangs vertically below the pulley.

What is the initial acceleration of B, and what is the tension in the string:

(*a*) if the table is smooth?

(*b*) if there is a constant friction force of 10 N between the table and particle A?

4 A railway locomotive of mass 55 tonnes is pulling a train of two trucks, each of mass 45 tonnes, along a level track. Each truck exerts a resistance to motion of 40 000 N.

(*a*) If the train is travelling at a constant speed, find:

(*i*) the driving force P(N) exerted by the locomotive,

(*ii*) the tension in the coupling between the two trucks, and the tension in the coupling between the locomotive and the first truck.

(*b*) The coupling between the two trucks now breaks.

Find:

(*i*) the initial acceleration of the locomotive and the remaining truck,

(*ii*) the new tension in the coupling,

(*iii*)the initial deceleration of the truck that has broken free from the train.

5 Two particles A and B, of mass 5 kg and 6 kg respectively, are connected by a light inextensible string that passes over a smooth pulley. The two particles are held stationary below the pulley, with the string just taut, and then released.

(*a*) Find the acceleration of the particles.

(*b*) Find the tension in the string.

3 Equilibrium

Unit 2 (Forces and Newton's laws) dealt with **dynamics**, with questions calculating how forces cause acceleration. This unit is about **statics**: how particles are kept in equilibrium by systems of forces.

> A consequence of Newton's second law
>
> $$F = ma$$
>
> is that **if a particle is in equilibrium, then the resultant of all the external forces must be zero**.

Equilibrium and force diagrams

For any particle in equilibrium, a vector diagram showing the sum of all the external forces must form a closed polygon, so that the force vectors sum to zero – the resultant vector is zero.

Particles in equilibrium under a system of exactly three external forces form a special case, because the force diagram is a triangle, and it can be analysed easily with the sine or cosine rules.

EXAMPLE

A cable car of mass 2 tonnes is suspended in equilibrium by two cables, one inclined at 45° to the vertical, and one at 30° on the opposite side of the vertical.

Find the tensions in the cables.

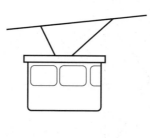 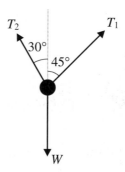 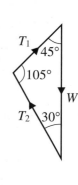

Using the sine rule in the diagram,

$$\frac{T_1}{\sin 30°} = \frac{T_2}{\sin 45°} = \frac{W}{\sin 105°}$$

$$\Rightarrow \quad T_1 = \frac{2000 \times 9.8 \times \sin 30°}{\sin 105°}$$

$$= 10\,146\,\text{N}$$

$$T_2 = \frac{2000 \times 9.8 \times \sin 45°}{\sin 105°}$$

$$= 14\,348\,\text{N}$$

So the tension in the cable inclined at 45° is 10 146 N, and in the other cable is 14 348 N.

Equilibrium and force diagrams (cont.)

EXAMPLE

A particle of mass 5 kg is suspended by a string. The string is held at an angle of 30° to the vertical by applying a force F N at 85° to the vertical.

Find the value of F.

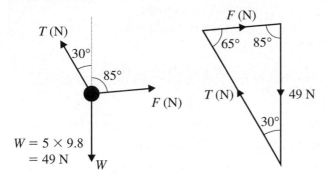

In the diagram,

$$\frac{F}{\sin 30°} = \frac{49}{\sin 65°} \Rightarrow F = \frac{49 \times \sin 30°}{\sin 65°} = 27.0 \text{ N} \quad (3 \text{ s.f.})$$

NOTE

Again here, and in the example below, we use T for the tension in the supporting string, and $g = 9.8 \text{ m s}^{-2}$.
If you work it out, $T = 53.9$ N.

EXAMPLE

The same particle as in the example above is still held in equilibrium with the suspending string at 30° to the vertical, but the extra force F N applied to the particle is now chosen so that it is a minimum.

Find the force F, and its angle to the vertical.

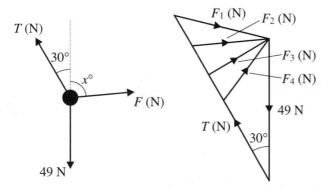

The triangle of forces could be drawn by taking any of the forces marked as F_1, F_2, ... in the diagram, though each will cause a different tension T in the suspending string.

But by inspection, the diagram which will generate the smallest possible magnitude for F is one in which F and T are perpendicular, as in the diagram opposite.

From the diagram,

$$F_{\text{min}} = 49 \sin 30°$$
$$= 24.5 \text{ N} \quad (3 \text{ s.f.})$$

and angle $x°$ is 60° to the vertical.

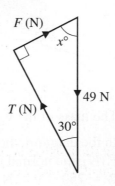

METHOD NOTE

You don't need the sine rule here – it's a right-angled triangle.

NOTE

Notice that this does not correspond to the smallest possible value of the tension T. Look at the triangle formed for the case F_4 in the diagram above.

General result for three forces

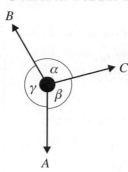

If a particle is acted on by three forces A, B and C in Newtons, in directions at angles α, β, γ apart as in the diagram on the left, the forces triangle diagram is as shown.

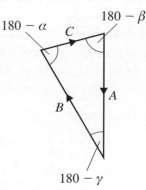

Using the sine rule in the forces triangle diagram gives:

$$\frac{A}{\sin(180° - \alpha)} = \frac{B}{\sin(180° - \beta)} = \frac{C}{\sin(180° - \gamma)}$$

and since, for any angle $x°$, $\sin(180 - x)° = \sin x°$, this is

$$\frac{A}{\sin \alpha} = \frac{B}{\sin \beta} = \frac{C}{\sin \gamma}$$

REVISION NOTE

This rule can easily be obtained in any question by drawing the forces triangle. If you prefer to have a rule, this one is sometimes called Lami's theorem, but it's really just the sine rule.

A particle in equilibrium on an inclined plane

A common type of **statics** question involves analysing the forces acting on a particle on an inclined plane. The important thing to remember is:

For a particle on an inclined plane:

the weight will always act vertically downwards, but the normal reaction force acting on the particle will be perpendicular to the plane.

REVISION NOTE

For more on friction see Unit 4, page 144.

EXAMPLE

A van of mass 0.75 tonnes is parked on a slope which is at 10° to the horizontal.

Find the friction force between the tyres and the road, and the normal reaction force acting on the van.

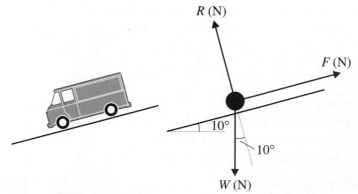

The mass of the van is 750 kg.

In the diagram, the friction force between the road and van is F (N), the normal reaction is R (N), and the weight is $W = 750 \times 9.8$ (N).

METHOD NOTE

Again we use $g = 9.8 \, \text{m s}^{-2}$, and treat the van as a particle.

NOTE

The friction force acts **up** the plane to prevent the van rolling down.

A particle in equilibrium on an inclined plane (cont.)

We **resolve** the forces parallel and perpendicular to the slope.

Since the van is in equilibrium (and so there is no acceleration), we just equate the components of the forces in opposite directions.

Using this diagram of the components of W and the forces diagram on page 140:

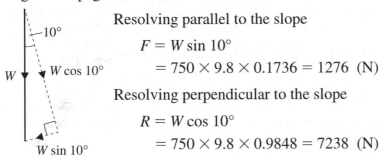

Resolving parallel to the slope

$$F = W \sin 10°$$
$$= 750 \times 9.8 \times 0.1736 = 1276 \ (N)$$

Resolving perpendicular to the slope

$$R = W \cos 10°$$
$$= 750 \times 9.8 \times 0.9848 = 7238 \ (N)$$

So the friction force between the tyres and the road is 1276 N up the slope, and the normal reaction force acting on the van is 7238 N.

REVISION NOTE

Remember that '*resolving*' means separating a force into two **components** at right angles to each other.

METHOD NOTE

Resolving parallel and perpendicular keeps F and R 'separate'. If, instead, you resolve the forces horizontally and vertically you can still solve the problem, but you will get a pair of simultaneous equations, each containing both F and R. So the choice is important.

EXAMPLE

A man is dragging a packing case up a slope of 30° at constant speed. The rope makes an angle of 20° with the slope. The friction force between the case and the slope is 200 N, and the mass of the case is 100 kg. Find the tension in the rope.

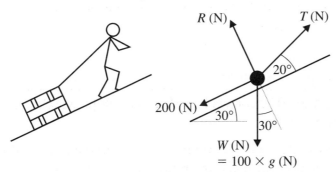

In the diagram there are two unknown forces, T and R. Since the question does not ask for R, it makes sense to resolve parallel to the slope. R is perpendicular to the slope, so its component parallel to the slope is zero.

So, resolving parallel to the slope,

$$T \cos 20° = F + W \sin 30°$$
$$= 200 + 100 \times 9.8 \times 0.5$$
$$= 690$$
$$\Rightarrow \quad T = \frac{690}{0.9397}$$
$$= 734 \ (N) \ (3 \ s.f.)$$

So the tension in the rope is 734 Newtons.

METHOD NOTE

There is a very important point here. Because the case is being dragged *at constant speed*, there is no acceleration. So this problem must be dealt with in exactly the same way as the **statics** problems – the resultant force resolved in any direction equals zero.

METHOD NOTE

Here we have used T (N) for the tension in the rope, and R (N) for the normal reaction from the slope on the case. We have also used $g = 9.8 \ m \ s^{-2}$.

EXAM NOTE

If the question asks for the normal reaction force, find it easily by resolving perpendicular to the slope.

Equilibrium problems using vector methods

The type of **statics** problems on pages 140 and 141 could also be solved using vector methods, but when all the forces act in two dimensions vector methods are usually unnecessary.

The following example is in three dimensions, and vectors make the solution easy.

REVISION NOTE

For a 2-D question, this vector approach will probably not be simpler than 'resolving forces'. But the methods are equivalent, so it's your choice.

EXAMPLE

The diagram shows a radio transmitter mast held in place in a vertical position by three cables anchored in the horizontal ground. The mast is 30 m tall.

Using a set of 3-D coordinates as indicated on the diagram, with origin O at the foot of the mast, the cables are anchored at $(-20, 0, 0)$, $(30, 15, 0)$, $(30, -15, 0)$, and the top of the mast $(0, 0, 30)$. Two of the cables have tension 700 N as shown.

Find the tension in the third cable, and the thrust in the radio mast.

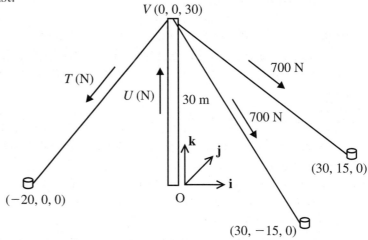

METHOD NOTE

For the direction of each cable tension, a column vector is the displacement vector from the top of the mast to the anchor point at ground level. So for T the direction is $-20\mathbf{i} + 0\mathbf{j} - 30\mathbf{k}$.

REVISION NOTE

Unit vectors and the magnitude of a vector are explained on page 121.

In this problem, there are four forces acting on the point V at the top of the mast: the tensions in the 3 cables, and the thrust, U.

These four forces can each be written as a column vector, with $\mathbf{i}$, $\mathbf{j}$, $\mathbf{k}$ components.

First write each of the forces as a vector. As an example, for the cable fixed at $(30, 15, 0)$ a unit vector in the direction of the tension can be written as:

METHOD NOTE

Each force can be written by multiplying its magnitude by a unit vector in the correct direction.

So the vector representing the tension in this cable is 700 times this unit vector.

$$\frac{1}{\sqrt{30^2 + 15^2 + 30^2}}\begin{pmatrix} 30 \\ 15 \\ -30 \end{pmatrix} = \frac{1}{45}\begin{pmatrix} 30 \\ 15 \\ -30 \end{pmatrix}$$

So, for the four forces at point V,

$$\frac{700}{45}\begin{pmatrix} 30 \\ 15 \\ -30 \end{pmatrix} + \frac{700}{45}\begin{pmatrix} 30 \\ -15 \\ -30 \end{pmatrix} + \frac{T}{\sqrt{1300}}\begin{pmatrix} -20 \\ 0 \\ -30 \end{pmatrix} + \frac{U}{30}\begin{pmatrix} 0 \\ 0 \\ 30 \end{pmatrix} = \begin{pmatrix} 0 \\ 0 \\ 0 \end{pmatrix}$$

METHOD NOTE

Because the mast is in equilibrium, the vector *resultant* of the four forces must be zero.

The top and bottom rows of this vector equation give

$$466\tfrac{2}{3} + 466\tfrac{2}{3} - 0.5547T = 0 \implies T = \frac{932.8}{0.5547} = 1680 \text{ N} \quad (3 \text{ s.f.})$$

$$-466\tfrac{2}{3} - 466\tfrac{2}{3} - 0.8321T + U = 0 \implies U = 2330 \text{ N} \quad (3 \text{ s.f.})$$

METHOD NOTE

We use the top row and the bottom row because these lead directly to T and U. The middle row involves neither T nor U.

Equilibrium

(Use $g = 9.8 \text{ m s}^{-2}$ where appropriate.)

1   A circus artist of mass 60 kg is suspended during a 'flying' act by two cables inclined at 30° and 45° to the vertical.

Find the tensions in the two cables.

2   A loaded sledge of mass 130 kg is at rest on an icy slope at an angle of 15° to the horizontal.

(a)   The friction force between the sledge and the slope is just sufficient to prevent the sledge beginning to slide down the slope.

Find the friction force.

(b)   A dog-pack now applies a force parallel to, and directly up, the slope.

How great a force must be applied by the dog-pack before the sledge begins to move up the slope?

3   A man pulls a trunk at constant speed up a slope which is at 20° to the horizontal. The trunk has mass 70 kg and the friction force between the trunk and slope is a constant 150 N. The rope used by the man to pull the trunk makes an angle of 30° to the slope, directly up the slope.

(a)   Find the tension in the rope.

(b)   What is the normal reaction between the trunk and the slope?

4   A chandelier is suspended from the roof of a large dining hall by 3 cables. The forces in two of the cables can be represented by $2\mathbf{i} + \mathbf{j} + 5\mathbf{k}$ and $2\mathbf{i} - \mathbf{j} + 5\mathbf{k}$. The weight of the chandelier is 15 Newtons.

What is the force in the third supporting cable? Calculate its magnitude.

5   A car of mass 0.7 tonnes is parked facing directly downhill on a slope of 30° to the horizontal.

If the handbrake holds the car stationary, find:

(a)   the normal contact force between the car and the slope,

(b)   the friction force acting up the slope.

6   The handbrake on the car in question **5** is old and weak, and will fail if the required friction force between the car and the slope is greater than 1500 N.

What is the angle to the horizontal of the steepest slope on which the car can be parked facing directly downhill without the brake failing?

7   A space craft in deep space is acted upon by three forces, which are given in terms of $\mathbf{i}, \mathbf{j}$ and $\mathbf{k}$, unit vectors in perpendicular directions relative to the space craft.

These forces are

$$5\mathbf{i} - b\mathbf{j} + 6\mathbf{k} \qquad -2\mathbf{i} + 3\mathbf{j} - c\mathbf{k} \qquad a\mathbf{i} + \mathbf{j} + 4\mathbf{k},$$

each measured in Newtons. The space craft is not accelerating.

Find the values of a, b and c.

4 Friction

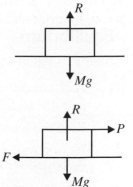

When a block of mass M kg rests on a horizontal surface, the forces acting on the block are **equal and opposite** as the block is in **equilibrium**.

Suppose a small horizontal force P (N) is applied to the block. The block may not move, because this force is opposed by the **friction** force F (N) acting between the block and surface.

> The block will not move while $F = P$.

The magnitude of F depends upon the 'roughness' of the contact surfaces. **As P increases, F also increases until it can get no larger**.

When $P > F$, the block will move according to Newton's second law

$$P - F = Ma$$

where a m s^{-2} is the acceleration of the block.

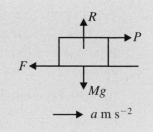

The coefficient of friction, μ

Maximum friction is given by

$$F = \mu R$$

where R is the reaction force of the body on the surface and μ is the coefficient of friction between the two surfaces.

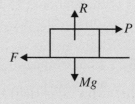

EXAMPLE

A block of mass 600 g rests on a surface where $\mu = 0.1$. A horizontal force of 1 N is applied to the block.

(a) Calculate the friction between the block and the surface.

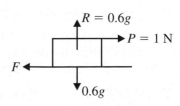

(b) Does the block move? If so, find the acceleration.

(a) Max $F = 0.1 \times 0.6g$

$\qquad = 0.588$

(b) Since $P > F$ the block will move and $P - F = ma$ so

$\qquad 1 - 0.588 = 0.6 \times a$

therefore $a = 0.69$ m s^{-2}

Forces applied at an angle

If a force P, inclined at an angle $\theta°$, acts on a block which sits on a rough horizontal surface, there are two effects.

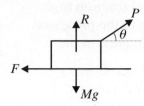

1   The horizontal component of P acting is $P \cos \theta$ and so for movement $F < P \cos \theta$.

2   The reaction force R is reduced because P has a vertical component $P \sin \theta$

and so $P \sin \theta + R = Mg$

Because R is reduced it follows that μR is also reduced and hence a smaller horizontal force is required to move the block.

NOTE

$P \cos \theta$ is the only force that can move the block.

METHOD NOTE

Resolving vertically.

EXAM NOTE

Be careful – in an exam it is easy to forget that R is reduced by the force at an angle.

EXAMPLE

A 10 kg box rests on a rough horizontal surface. The coefficient of friction is 0.40.

(a) If a force $P = 38$ N is applied horizontally, will the box move?

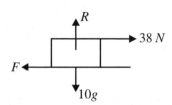

(b) If the force P is now applied at an angle of 10° to the horizontal, will the box move?

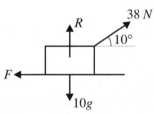

(a) Max $F = 0.40 \times 10g = 39.2$ N

The force applied is only 38 N, so the box will not move.

(b) $R = 10g - 38 \sin 10°$

         $= 91.4$ N

Max $F = 0.40 \times 91.4 = 36.56$ N

Horizontal component of the force $= 38 \cos 10°$

                $= 37.42$ N

Max $F <$ horizontal component of the force

Therefore the box will move.

METHOD NOTE

Resolving vertically:
$R + 38 \sin 10° = 10g$ where $g = 9.8 \text{ m s}^{-2}$.

Friction on rough inclined planes

If a mass M kg sits on a rough plane inclined at an angle of $\theta°$ to the horizontal, the weight Mg acts vertically, the normal reaction to the surface R acts at right angles to the plane and friction F acts up the plane to prevent the block from slipping.

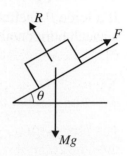

Resolving the forces

Forces must be resolved **parallel to the plane and perpendicular to the plane**.

Parallel to the plane:
$$F = Mg \sin \theta$$

Perpendicular to the plane:
$$R = Mg \cos \theta \text{ and}$$

Max $F = \mu Mg \cos \theta$

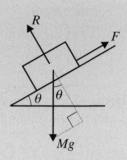

METHOD NOTE

Forces must be resolved in two perpendicular directions. You could choose vertical and horizontal directions but they would not be very helpful here. *Remember* – you should try to choose directions that eliminate some of the unknown forces.

When the block is about to slip:
$$Mg \sin \theta = \mu Mg \cos \theta$$

For the block to slide down the plane:
$$Mg \sin \theta > \mu Mg \cos \theta$$

Hence $\tan \theta > \mu$

The equation of motion in this case is
$$Mg \sin \theta - F = Ma$$

METHOD NOTE

Cancel Mg and divide by $\cos \theta$.

REVISION NOTE

a is the acceleration down the plane.

EXAMPLE

A block of mass 1.5 kg sits on a rough plane, inclined at 15° to the horizontal. The coefficient of friction is 0.2.

Calculate the reaction force and determine whether or not the block will slide down the plane.

Resolving perpendicular to the plane:
$$R = 1.5g \cos 15° = 14.2 \text{ N}$$

Force down the plane is:
$$1.5g \sin 15° = 3.8 \text{ N}$$

Max $F = 0.2 \times R = 2.84 \text{ N}$

Therefore, the box will slide down the plane.

The equation of motion is given by

Force down the plane − friction = mass × acceleration
$$3.8 - 2.84 = 1.5 \times a$$

and so $a = 0.64 \text{ m s}^{-2}$

$g = 9.8 \text{ m s}^{-2}$

METHOD NOTE

The box slides down the plane because

  Max $F <$ force down the plane.

METHOD NOTE

Newton's second law
  $F = ma$

Motion up an inclined plane

A block of mass M kg rests on a rough plane, inclined to the horizontal at an angle $\theta°$. The coefficient of friction is μ.

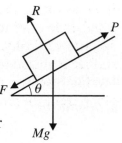

A force P (N) is applied to the block in a direction parallel to the plane as shown.

The block will move up the plane if P is greater than the combined forces of friction and the component of the weight parallel to the plane.

Resolving the forces

Perpendicular to the plane:

$R = Mg \cos \theta$

and $F = \mu \, Mg \cos \theta$

Parallel to the plane:

$F + Mg \sin \theta$ (down the plane)

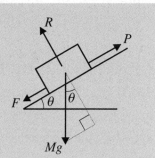

METHOD NOTE

Using $F = \mu R$

The equation of motion is:

$P - (F + Mg \sin \theta) = Ma$

So if $P > \mu Mg \cos \theta + Mg \sin \theta$ the block will move up the plane.

METHOD NOTE

Using Newton's second law $F = ma$, where a is the acceleration.

EXAMPLE

A block of mass 2.5 kg rests on a rough plane inclined at $10°$ to the horizontal. The coefficient of friction is $\mu = 0.3$.

Calculate to 2 d.p. the force needed to move the block up the plane with an acceleration of 0.1 m s^{-2}.

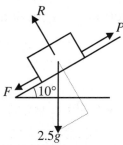

Resolving perpendicular to the plane:

$R = 2.5g \cos 10°$

and so $F = 0.3 \times 2.5g \cos 10°$

$= 7.238 \text{ N}$

The equation of motion is:

$P - (F + Mg \sin 10°) = Ma$

$P - (7.238 + 4.248) = 2.5 \times 0.1$

$P = 11.74 \text{ N}$

REVISION NOTE

Friction acts *down* the plane because the block is being moved *up* the plane.

METHOD NOTE

Newton's second law:

$F = ma$

The LHS is 'forces up the plane − forces down the plane'.

Friction

(Use $g = 9.8 \text{ m s}^{-2}$ where appropriate.)

1   For the system shown, calculate the friction.

State whether the body will accelerate or
remain at rest, given that $\mu = \frac{1}{8}$.

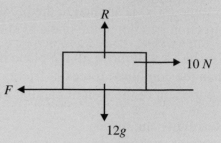

2   A body of mass 1.5 kg is at rest on a rough horizontal surface where the coefficient of
friction is 0.2.

A constant horizontal force, T is applied for 10 seconds and then removed.

Given that the body had a velocity of 2.5 m s^{-1} when
the force was removed, find:

(*a*)   the acceleration of the body while the force was acting,

(*b*)   the magnitude of the applied force,

(*c*)   the retardation of the body when the force is removed,

(*d*)   the total distance moved by the body.

> **HINT**
> (*a*) Use $v = u + at$
> (*b*) Use $F = ma$
> (*c*) Use $F = ma$
> (*d*) Use $v^2 = u^2 + 2as$

3   For the system shown, calculate R, the reaction to the
plane.
Will the block accelerate along the surface,
given that $\mu = 0.3$?
Give reasons.

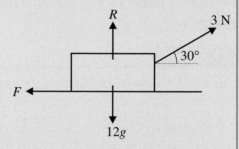

4   A block of mass 1 kg rests in equilibrium on a rough horizontal table with coefficient of
friction μ. A force P acts at an angle of 25° to the horizontal.
Given that $P = 3 \text{ N}$ calculate:

(*a*)   the normal reaction exerted by the table on the block,

(*b*)   the frictional force on the block.

(*c*)   Given that the block is about to move,
calculate the coefficient of friction to 2 d.p.

> **HINT**
> Draw a diagram showing all the
> forces acting on the block.

5   For the system shown $\mu = 0.5$.
Find the magnitude of the friction and state
whether the body will remain at rest.

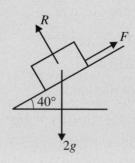

5 Momentum and impulse

Momentum

> The **momentum** of a body is given by mv, where m is the mass in kg and v is the velocity in m s^{-1}.
>
> The units of momentum are Ns (Newton seconds).

When the velocity of a body changes, its momentum also changes. This change in momentum is given by

$$mv - mu$$

where u is the initial velocity and v is the final velocity of the body.

The change in velocity of the body is the result of a force acting upon the body.

REVISION NOTE

This form of the result is only true for motion in a straight line – which is all you need for AS-Level.

EXAMPLE

Find the change in momentum when a body of mass 3 kg changes speed from 2 m s^{-1} to 7 m s^{-1}.

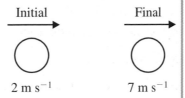

Initial   Final

2 m s^{-1} 7 m s^{-1}

Change in momentum:

$mv - mu = 3 \times 7 - 3 \times 2 = 15$ Ns

METHOD NOTE

Always draw a before and after diagram.

Impulse

> **Impulse** is given by $F \times t$, where F is a constant force and t is the time for which the force acts.

We know that $F = ma$, so impulse $= ma \times t$.

By substituting $t = \dfrac{v - u}{a}$, we find that

$$\text{Impulse} = m\frac{(v - u)}{a} \times a = mv - mu$$
$$= \text{change in momentum.}$$

The units of impulse are Ns (Newton seconds).

REVISION NOTE

Force is in Newtons or kg m s^{-2} and impulse is in Newton seconds, so

$$\text{kg m s}^{-1} = \text{kg m s}^{-2} \times \text{s}$$
$$= F \times t$$

From $v = u + at$

EXAMPLE

A force of 5 N acts on a body of mass 4 kg for 8 seconds. Find the final velocity if the initial velocity was 3 m s^{-1}.

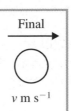

Initial   Final

3 m s^{-1} v m s^{-1}

Impulse $= F \times t = 5 \times 8 = 40$ Ns

Change in momentum $= mv - mu$

$\qquad\qquad\quad = 4v - 4 \times 3 = $ Impulse

so $\quad 40 + 12 = 4v$

therefore $\quad v = 13$ m s^{-1}

Collisions

Two bodies moving in a straight line at constant velocities have momenta m_1u_1 and m_2u_2.

If a collision between the two bodies occurs then although kinetic energy will be lost, the total momentum of the system is conserved.

The bodies will continue after the collision with momenta m_1v_1 and m_2v_2 such that

$$m_1u_1 + m_2u_2 = m_1v_1 + m_2v_2$$

REVISION NOTE

This is known as the **Principle of Conservation of Linear Momentum**.

REVISION NOTE

Momentum before impact = momentum after impact.

EXAMPLE

A body of mass 1 kg is moving on a smooth horizontal surface at $2\,\text{m s}^{-1}$ when it collides with a stationary body of mass 3 kg. The larger body moves away at $1\,\text{m s}^{-1}$.

How does the smaller body move after impact?

Taking the positive direction to the right we have these before and after diagrams.

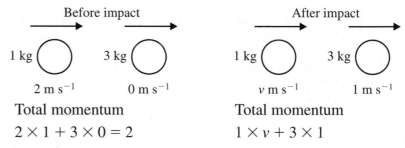

Total momentum

$2 \times 1 + 3 \times 0 = 2$

Total momentum

$1 \times v + 3 \times 1$

The momentum before impact must be equal to the momentum after impact, so

$$2 = v + 3$$

and

$$v = -1\,\text{m s}^{-1}$$

The 1 kg body moves to the left at $1\,\text{m s}^{-1}$.

EXAM NOTE

Always draw a before and after diagram to clarify the problem.

METHOD NOTE

$v = 1\,\text{m s}^{-1}$

The negative velocity means it is moving in the opposite direction.

EXAMPLE

A body of mass 1 kg is moving on a smooth horizontal surface at $2\,\text{m s}^{-1}$ when it collides with a stationary body of mass 3 kg. The two bodies combine and move away together.

Calculate the velocity with which they move.

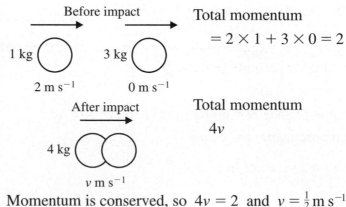

Total momentum

$= 2 \times 1 + 3 \times 0 = 2$

Total momentum

$4v$

Momentum is conserved, so $4v = 2$ and $v = \frac{1}{2}\text{m s}^{-1}$

Momentum and impulse

1   A body of mass 5 kg is moving with a constant velocity of 3 m s^{-1}. After a force P N has acted on the body for 2 seconds it moves with a velocity of 5 m s^{-1}. Calculate P.

HINT
Draw before and after diagrams.

2   A body of mass 3 kg is moving with a velocity of u m s^{-1}. After a force of 2 N acts on it for 3 seconds the body moves with a velocity of 5 m s^{-1}.

What was the initial velocity?

REMEMBER
$Ft = mv - mu$

3   A block of wood of mass 1.5 kg is stationary on a smooth horizontal surface when a bullet of mass 0.15 kg hits it. The bullet is travelling horizontally with a speed of 385 m s^{-1} when it hits the wood. The bullet does not leave the wood.

Calculate the speed of the block of wood after the impact.

HINT
All the momentum from the bullet has gone into the movement of the bullet and the block of wood.

4   The following diagrams show the before and after conditions of two bodies in collision. Calculate the value of v in both cases.

(*a*)

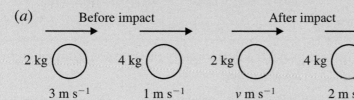

(*b*)

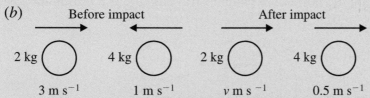

5   A toy car of mass 0.25 kg is travelling in a straight line at a speed of 0.3 m s^{-1} when it collides with a smooth wooden block of mass 0.1 kg. The car and the block then move along together at a constant speed.

(*a*) Find the speed with which the car and block are moving after the collision.

(*b*) Find the impulse exerted on the block by the car.

HINT
The momentum of the car has gone into the momentum of the car and block combined.

6   A body of mass 5 kg is initially at rest on a smooth horizontal surface. A horizontal force of 7 N acts on the body for 5 seconds.
Find:

(*a*) the magnitude of the impulse given to the body,

(*b*) the magnitude of the final momentum of the body,

(*c*) the final speed of the body.

6 Projectiles

Horizontal projection

When an object is projected in a horizontal direction from a position above level ground its motion can be considered in two parts: vertical motion and horizontal motion.

Vertical motion

In simple cases, the only force acting on the projectile in flight is gravity, because we model the projectile as a particle with mass but no volume. As gravity acts vertically downwards, it will only affect the vertical component of the projectile's motion. The gravitational force accelerates the particle in the downward direction.

Horizontal motion

There is no force acting in the horizontal direction after the initial projection force, so the particle will travel at a constant horizontal speed throughout its flight.

You can use the constant acceleration formulae, $v = u + at$, $v^2 = u^2 + 2as$ and $s = ut + \frac{1}{2}at^2$ to solve the problem.

REVISION NOTE

We are assuming *no air resistance* – this is another example of **modelling**.

EXAM NOTE

Always draw a diagram.

METHOD NOTE

Always write down all the initial conditions. Consider the vertical and horizontal motion separately.

EXAM NOTE

Always use $g = 9.8\,\text{m s}^{-2}$ unless the question tells you otherwise, as in this case.

EXAMPLE

A particle is projected horizontally at $10\,\text{m s}^{-1}$ from a height of $180\,\text{m}$ above level ground.

Find the time taken by the particle to reach the ground and find the horizontal distance travelled before hitting the ground. ($g = 10\,\text{m s}^{-2}$)

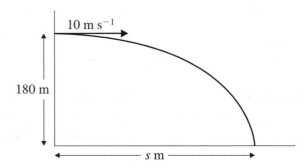

Vertical motion	Horizontal motion
$u = 0\,\text{m s}^{-1}$	$u = 10\,\text{m s}^{-1}$
$s = 180\,\text{m}$	$s = ?\,\text{m}$
$t = ?\,\text{s}$	$t = ?\,\text{s}$
$a = g = 10\,\text{m s}^{-2}$	$a = 0\,\text{m s}^{-2}$

Vertical motion:
$$s = ut + \tfrac{1}{2}at^2$$
$$180 = \tfrac{1}{2} \times 10 \times t^2$$
$$36 = t^2$$
$$t = 6\,\text{s}$$

Horizontal motion:
$$s = ut + \tfrac{1}{2}at^2$$
$$s = 10 \times 6 + \tfrac{1}{2} \times 0 \times 6^2$$
$$s = 60\,\text{m}$$

Therefore the time taken to hit the ground is 6 seconds. In this time the particle travels a distance of 60 metres in the horizontal direction.

Projectiles launched at an angle to the horizontal

When a particle is projected at an angle of θ above a horizontal plane with velocity u m s^{-1}, the motion should again be considered in two parts:

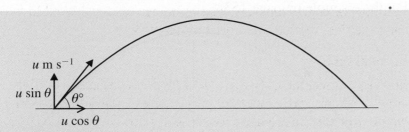

Vertical motion

Velocity $= u \sin \theta$ m s^{-1}

Acceleration $= -9.8$ m s^{-2}

Horizontal motion

Velocity $= u \cos \theta$ m s^{-1}

Acceleration $= 0$ m s^{-2}

REVISION NOTE

The upwards direction is taken to be positive. So acceleration g, which acts downwards, is negative.

EXAMPLE

A small stone is projected into the air with a velocity of 10 m s^{-1} at an angle of 50° from a horizontal surface.

Calculate the time taken to reach its maximum height and the horizontal distance travelled in that time.

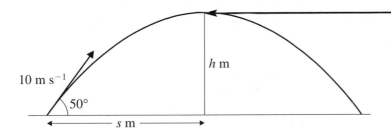

REVISION NOTE

At this point vertical velocity is zero, as the object comes to rest before starting to fall.
See page 156.

EXAM NOTE

Draw a diagram to help you understand what is happening.

Vertical motion

Velocity $= 10 \sin 50°$ m s^{-1}

Acceleration $= -9.8$ m s^{-2}

$t = ?$

$h = ?$

Horizontal motion

Velocity $= 10 \cos 50°$ m s^{-1}

Acceleration $= 0$ m s^{-2}

$t = ?$

$s = ?$

REMEMBER

You can use all your **constant acceleration** formulae.

Velocity at max height $= 0$ m s^{-1}

Using $v = u + at$

$0 = 10 \sin 50° - 9.8t$

$t = 0.781\,678$ s

Using $v^2 = u^2 + 2ah$

$0 = (10 \sin 50°)^2 - 2 \times 9.8h$

$h = 2.99$ m (2 d.p.)

Velocity is constant throughout the flight.

Using $s = ut + \frac{1}{2}at^2$

$s = 10 \cos 50° \times 0.781\,678$

$s = 5.02$ m (2 d.p.)

EXAM NOTE

Do not round off too early as this could cause errors in the final answer.

Therefore the particle reaches a height of approximately 3 m in 0.78 seconds, and during this time it travels 5 m in the horizontal direction.

Symmetry of the path of a projectile

To calculate the horizontal distance a projectile travels before hitting the surface again, we can use the fact that the trajectory or path of the projectile is symmetrical, as shown below.

For the example on page 153:

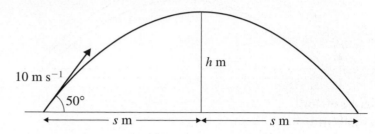

Total horizontal distance travelled is $2 \times 5.02 = 10.04$ m.

Total time of flight $= 2 \times 0.78 = 1.56$ s.

Projection at an angle from a point above the plane

When a particle is projected at an angle from a point above the horizontal plane, the information given will help to tell you which direction of motion to use first.

EXAMPLE

A stone is kicked from the edge of a vertical cliff with a velocity of 15 m s^{-1}, at an angle of 20° to the horizontal. The stone hits the ground 250 m from the base of the cliff.

Find the time that the stone is in the air and also the height of the cliff.

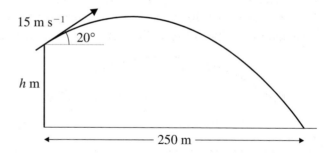

Vertical motion
$u = 15 \sin 20°$ m s^{-1}
$a = -9.8$ m s^{-2}
$t = ?$
$h = ?$

Horizontal motion
$u = 15 \cos 20°$ m s^{-1}
$a = 0$ m s^{-2}
$s = 250$ m
$t = ?$

Using $s = ut + \frac{1}{2}at^2$
$250 = 15 \cos 20° \times t$
$t = 17.7363$ s (4 d.p.)

Using $h = ut + \frac{1}{2}at^2$
$h = 15 \sin 20° \times 17.7363 + \frac{1}{2} \times -9.8 \times 17.7363^2$
$h = -1450$ m (to the nearest metre)

So the height of the cliff is 1450 m and the time taken to hit the ground is 17.74 s (2 d.p.).

Time of flight

Consider an object projected from a horizontal surface with an initial velocity of u m s^{-1} at an angle of $\theta°$ to the horizontal. It lands on the plane a distance d m from the point of projection.

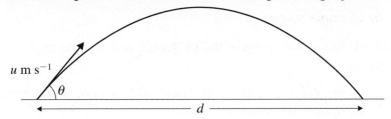

Vertical motion

$u = u \sin \theta$ m s^{-1}

$a = -g$ m s^{-2}

$h = 0$ when the object is on the plane.

Using $\quad s = ut + \frac{1}{2}at^2$

$\qquad 0 = u \sin \theta \times t - \frac{1}{2}gt^2$

Therefore

$\qquad 0 = t(u \sin \theta - \frac{1}{2}gt)$

Hence $\quad t = 0$ s or $t = \dfrac{2u \sin \theta}{g}$ s

Time of flight is given by: $\quad t = \dfrac{2u \sin \theta}{g}$ s

Range of a projectile

An object is projected from a horizontal plane with velocity u m s^{-1} at an angle θ. The range of the projectile is the horizontal distance covered between the point of projection and the point of landing.

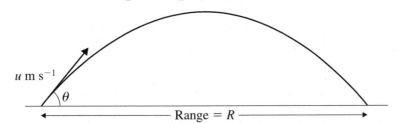

Vertical motion

$u = u \sin \theta$ m s^{-1}

$a = -g$ m s^{-2}

$t = \dfrac{2u \sin \theta}{g}$ s (time of flight)

Horizontal motion

$u = u \cos \theta$ m s^{-1}

$a = 0$ m s^{-2}

Using $s = ut + \frac{1}{2}at^2$

$R = u \cos \theta \times \dfrac{2u \sin \theta}{g}$

$R = \dfrac{u^2 \sin 2\theta}{g}$

This is true for all projectiles with the same assumptions.

Maximum range of a projectile

We found on page 155 that the range of a particle projected at an angle from a horizontal plane is given by

$$R = \frac{u^2 \sin 2\theta}{g}$$

where u is the initial speed of the projectile and θ is the angle of projection.

For any given value of u m s^{-1}, the range R m will be a maximum when $\sin 2\theta$ takes its maximum value of 1.

This occurs when $2\theta = 90°$, that is when $\theta = 45°$.

If $\theta = 45°$, the range $R = \dfrac{u^2}{g}$ m.

This is the maximum range of a projectile which has an initial speed of u m s^{-1}. For a particle to reach its maximum range its angle of projection must be 45°.

REVISION NOTE

This is true for all projectiles that start and finish their flights on the same horizontal plane, again with the same modelling assumptions.

Maximum height of a projectile

Considering only the vertical motion of the projectile, we can see that when it reaches its greatest height, the vertical component of the velocity is 0 m s^{-1}.

Vertical motion

$u = \sin \theta$ m s^{-1} $\qquad a = -g$ m s^{-2} $\qquad s =$ height (m).

Using $v^2 = u^2 + 2as$

$$0 = u^2 \sin^2 \theta - 2gs$$

Rearranging gives $s = \dfrac{u^2 \sin^2 \theta}{2g}$ m

This is the maximum height reached for a projectile with a given initial velocity of u m s^{-1} at an angle of θ.

METHOD NOTE

You can substitute the initial values into these equations when necessary.

EXAMPLE

A football is kicked into the air from ground level, on horizontal ground. Its initial velocity is 45 m s^{-1} at an angle of elevation of 35°.

Calculate the time of flight and the range for the ball.

Using the formulae:

$$t = \frac{2u \sin \theta}{g} \qquad\qquad R = \frac{u^2 \sin 2\theta}{g}$$

$$= \frac{2 \times 45 \sin 35°}{9.8} \qquad R = \frac{45^2 \sin 70°}{9.8} = 194.2 \text{ m}$$

$$t = 5.27 \text{ s}$$

REVISION NOTE

Total time of flight is 5.27 s and range of flight of ball is 194.2 m.

The maximum height reached is 33.99 m, and you could easily find this from the formula above if the question asked for it.

Equation of the trajectory

A particle is projected from a horizontal plane with an initial speed of u m s^{-1} at an angle of θ.

It follows a curved path, moving upward until it reaches its maximum height where the vertical component of its velocity $= 0$ m s^{-1}.

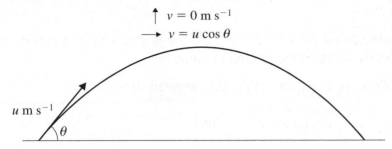

The particle then accelerates back towards the plane, following a curved path which is the reflection of the upward path.

During the time of flight the horizontal velocity remains constant at $u \cos \theta$ m s^{-1}. To find the cartesian equation of the path of the projectile (the trajectory), draw in the x and y axes so that the origin O is at the point of projection.

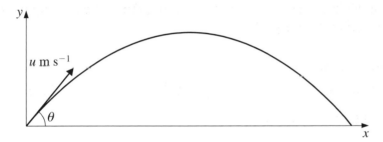

Vertical motion

$u = u \sin \theta$ m s^{-1}

$a = -g$ m s^{-2}

$t = ?$

$s = y$

$y = u \sin \theta t - \frac{1}{2}gt^2$ ⟵

Horizontal motion

$u = u \cos \theta$ m s^{-1}

$a = 0$ m s^{-2}

$t = ?$

$s = x$

$x = u \cos \theta t$ ⟵

$$t = \frac{x}{u \cos \theta}$$

> **METHOD NOTE**
>
> Using $s = ut + \frac{1}{2}at^2$

Substituting the expression for t into the expression for y gives:

$$y = u \sin \theta \times \frac{x}{u \cos \theta} - \frac{1}{2}g \frac{x^2}{u^2 \cos^2 \theta}$$

$$y = x \tan \theta - \frac{gx^2}{2u^2}(1 + \tan^2 \theta)$$

> **REVISION NOTE**
>
> $\dfrac{\sin \theta}{\cos \theta} \equiv \tan \theta$
>
> $\dfrac{1}{\cos^2 \theta} \equiv \sec^2 \theta$
>
> $\sec^2 \theta \equiv 1 + \tan^2 \theta$

The variables are x, y, θ and u.

If three of the variables are known, the fourth variable can be found by substitution.

Projectiles

(Use $g = 9.8 \text{ m s}^{-2}$ where appropriate.)

1 A particle is projected at 15 m s^{-1} in a horizontal direction from a height of 20 m above level ground. Find:

(*a*) the time taken by the particle to reach the ground,

(*b*) the horizontal distance travelled during flight.

2 A particle is projected horizontally from a point 2 m above horizontal ground. The particle hits the ground at a point 30 m horizontally from the point of projection.

Find the initial speed of the particle.

> **HINT**
> Similar to question 1, but you need to work backwards.

3 A cannon has its barrel set at an angle of elevation of 18°. The cannon fires a ball with an initial speed of 60 m s^{-1}.

Find the total horizontal distance travelled by the cannon ball and the time of flight.

> **HINT**
> Use the formulae.

4 A cricket ball is thrown upwards at an angle of 16° from a height h m above level ground. Its initial speed is 25 m s^{-1}. It hits the base of the stumps which are 40 m away.

Calculate the initial height of the ball.

5 A ball is kicked with a speed of 28 m s^{-1} at an angle of 15° from a point on horizontal ground.

Calculate the range on the horizontal plane.

At what angle should the ball be kicked for maximum range at this speed?

> **HINT**
> Use the formulae.

6 Rapunzel is standing at the window in her tower when she sees a prince approaching below. To attract his attention she wraps a message around her hair brush and throws it from the window with a speed of 14 m s^{-1} at an angle of 60° to the horizontal. It lands at the prince's feet, a horizontal distance of 28 metres from the base of the tower.

Find the height of Rapunzel's window from the base of the tower.

7 A particle is projected from a horizontal plane with a speed of 30 m s^{-1} at an angle of 40° to the horizontal. It hits the ground at a point which is level with the point of projection.

Find the time of flight and the range on the horizontal plane.

8 A particle is projected with velocity 80 m s^{-1} from level ground. The angle of elevation of projection is 25°.

Find the equation of the trajectory and also the greatest height reached by the particle during the flight.

> **HINT**
> Use the formulae.

Answers and hints to solutions – Mechanics

Exercise 1: Distance, velocity and acceleration

1 $18\,\mathrm{m\,s^{-1}}$, $108\,\mathrm{m}$ $12.25\,\mathrm{m\,s^{-1}}$

2 $18.975\,\mathrm{m}$, $2.581\,\mathrm{s}$, $5.3\,\mathrm{m\,s^{-1}}$ downwards

3 The bus travels at constant speed for 44.17 seconds.

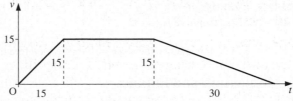

4 $\mathbf{v} = -3\mathbf{i} + 5\mathbf{j}$ so speed $= \sqrt{34}$, direction $121°$ to the positive x-axis.
$\mathbf{s} = +2\mathbf{j}$ so $\mathbf{r} = 3\mathbf{i} + 14\mathbf{j}$ and the particle is at $(3, 14)$.

5 (a) $s(t) = \frac{1}{2}t^2 + \frac{1}{3}t^3 + 2$

6 $33.3\,\mathrm{m}$

7 (a) $(0, 1)$, $(2, -2)$, $(8, -5)$, $(18, -8)$
(b) $\mathbf{v} = 4t\mathbf{i} - 3\mathbf{j}$, $\mathbf{a} = 4\mathbf{i}$
(c) $4\,\mathrm{m\,s^{-2}}$ parallel to x-axis
(d) velocity magnitude $5\,\mathrm{m\,s^{-1}}$, direction $36.9°$ below the x-axis.

Exercise 2: Forces and Newton's laws

1 $103\,\mathrm{N}$

2 If T is the towbar force, and D the drive force:
(a) $a = 1 \Rightarrow$ caravan: $T - 500 = 500 \Rightarrow T = 1000$ (tension)
   car & caravan: $D - 1000 = 2250 \Rightarrow D = 2250$
(b) $a = 0 \Rightarrow$ caravan: $T - 500 = 0 \Rightarrow T = 500$ (tension)
   car & caravan: $D - 1000 = 0 \Rightarrow D = 1000$
(c) $a = -1.5 \Rightarrow$ caravan: $T - 500 = -750 \Rightarrow T = -250$ (thrust)
   car & caravan: $D - 1000 = -1875 \Rightarrow D = -875$ (brake)

3 (a) $3.27\,\mathrm{m\,s^{-2}}$ downwards, $32.7\,\mathrm{N}$
(b) $2.6\,\mathrm{m\,s^{-2}}$ downwards, $36\,\mathrm{N}$

4 (a) (i) $80\,000\,\mathrm{N}$ (ii) $40\,000\,\mathrm{N}$, $80\,000\,\mathrm{N}$
(b) (i) $0.4\,\mathrm{m\,s^{-2}}$ (ii) $58\,000\,\mathrm{N}$ (iii) $-0.09\,\mathrm{m\,s^{-2}}$

5 (a) A $0.89\,\mathrm{m\,s^{-2}}$ upwards, B $0.89\,\mathrm{m\,s^{-2}}$ downwards
(b) $53.45\,\mathrm{N}$

Exercise 3: Equilibrium

1 $430\,\mathrm{N}$, $304\,\mathrm{N}$

2 (a) $330\,\mathrm{N}$ (b) $660\,\mathrm{N}$

3 (a) Resolve parallel to the slope:
$T \cos 30° = 150 + 686 \sin 20° \Rightarrow T = 444\,\mathrm{N}$
(b) Resolve perpendicular to the slope:
$T \sin 30° + R = 686 \cos 20° \Rightarrow R = 423\,\mathrm{N}$

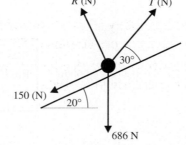

4 Force $-4\mathbf{i} + 5\mathbf{k}\,\mathrm{N}$, magnitude $= \sqrt{41} \approx 6.40\,\mathrm{N}$

5 (a) $5940\,\mathrm{N}$ (b) $3430\,\mathrm{N}$

6 $12.6°$

7 $a = -3, b = 4, c = 10$

Exercise 4: Friction

1 Max $F = 14.7$ N, therefore remains at rest.

2 (*a*) $a = 0.25$ m s^{-2} (*b*) $T = 3.315$ N   (*c*) $a = -1.96$ m s^{-2}
   (*d*) Distance moved = 14.1 m (12.5 m while force is acting, then 1.594 m while slowing down).

3 $R = 13.2$ N, Max $F = 3.96$ N, horizontal force = 2.60 N, therefore block does not move.

4 (*a*) $R = 8.532$ N   (*b*) Max $F = 8.532\mu$ N
   (*c*) $3 \cos 25° = 8.532\mu$, therefore $\mu = 0.32$

5 $R = 15.01$ N, $F = 7.51$ N, Force down the slope = 12.60 N, therefore the block slides down the slope.

Exercise 5: Momentum and impulse

1 $P = 5$ N (momentum before force applied = 15 Ns; momentum after the force = 25 Ns and $mv - mu = Ft$)

2 $u = 3$ m s^{-1}

3 $v = 35$ m s^{-1} (momentum of bullet = 0.15×385 Ns; momentum of bullet plus wood = $1.65 \times v$ Ns)

4 (*a*) $v = 1$ m s^{-1} (total momentum = 10 Ns)
   (*b*) $v = 0$ m s^{-1} (total momentum = 2 Ns)

5 (*a*) $v = 0.214$ m s^{-1} (total momentum = 0.075 Ns)
   (*b*) Impulse = -0.0214 Ns (the car loses momentum)

6 (*a*) 35 Ns   (*b*) 35 Ns   (*c*) 7 m s^{-1}

Exercise 6: Projectiles

1 (*a*) $t = 2.02$ s   (*b*) $s = 30.3$ m

2 $t = \dfrac{30}{u}$, $2 = \frac{1}{2} \times 9.8 \times \left(\dfrac{30}{u}\right)^2$, $u = 46.96$ m s^{-1}

3 $R = \dfrac{u^2 \sin 2\theta}{g}$, $R = 215.9$ m, $t = \dfrac{2u \sin \theta}{g}$, $t = 3.78$ s

4 **Horizontal motion**   **Vertical motion**
   $u = 25 \cos 16°$ m s^{-1} $u = 25 \sin 16°$ m s^{-1}
   $a = 0$ $a = -9.8$ m s^{-2}
   $d = 40$ m

   $t = \dfrac{40}{25 \cos 16°} = 1.66$

   $h = 25 \sin 16° \times 1.66 + \frac{1}{2} \times (-9.8) \times 1.66^2$
   $h = -2.106$ m. Therefore the initial height of the ball was 2.1 m.

5 $R = \dfrac{u^2 \sin 2\theta}{g}$, $R = 40$ m; maximum range comes from elevation of 45° and is 80 m.

6 $h = 29.9$ m

7 $t = \dfrac{2u \sin \theta}{g} = 3.94$ s, $R = 90.4$ m

8 $y = x \tan 25° - \dfrac{g \times x^2}{2 \times 80^2}(1 + \tan^2 25°)$; $y = 0.47x - 0.0009x^2$

> Max height = 58.32 m
> Range = 500.27 m
> Time = 6.9 s

Decision and Discrete Maths Topics:

1 Algorithms

Flow charts	162
Sorting algorithms	162
Bubble sort	162
Shuttle sort	163
Quick sort	164
Comparing sorting methods	164
Binary search	165
Bin packing	165
Full bin algorithm	165
First fit algorithm	166
First fit decreasing algorithm	166
Comparing solutions	166
Exercise 1: Algorithms	167

2 Graphs and networks

Vocabulary and definitions	168
Euler tour	169
Hamiltonian cycle	169
Planar and non-planar graphs	169
Odd vertices in a graph	169
Bipartite graphs and matchings	170
Finding a maximum matching	170
Minimal spanning trees	171
Prim's algorithm	171
Kruskal's algorithm	171
Flows in networks	172
Maximum flow	172

3 Shortest path problems

Dijkstra's shortest path algorithm	173
The Chinese postman problem (route inspection)	174
Exercise 2: Networks and shortest path problems	176

4 Critical path analysis

Precedence networks	177
Cascade charts	178
Resources histograms	179

5 Linear programming

Graphical solution using vertices	180
Linear programming and integer solutions	181
The Simplex method	182

6 Boolean algebra

Propositional logic symbols	184
Truth tables	184
Laws of Boolean algebra	185
De Morgan's laws	185
Switching circuits	186
Combinatorial circuits	187
Exercise 3: Miscellaneous questions	188

Answers and hints to solutions	188

1 Algorithms

An algorithm is a procedure or set of rules for completing complicated calculations or for solving a problem.

If an algorithm is written as a set of rules, the rules must be precise and must work for any set of input values. They must produce an answer or output and should stop after a finite number of steps.

To get full marks on any questions involving algorithms you must follow the rules exactly. A computer cannot think for itself, it just does what the rules tell it. You must do the same thing.

Flow charts

An algorithm may be written as a flow chart. This is particularly useful when a process must be repeated several times before an output is obtained.

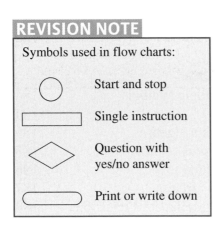
EXAMPLE

This flow chart will calculate all the cubes from 1^3 to the first cube after 500.

It will print out the number N and its cube S for all N for which $N^3 < 500$.

The printout will look like this:

     1, 1

     2, 8

     3, 27 …

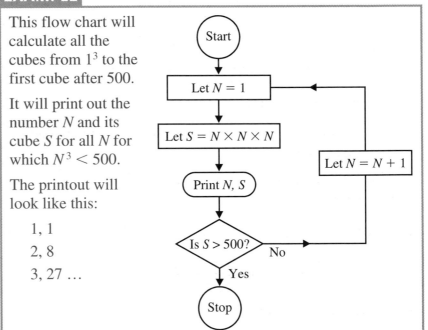

Sorting algorithms

These are for sorting numbers, names or items into ascending or descending order.

Bubble sort

1 Compare the 1st and 2nd numbers in the list. Swap if they are not in ascending order.

2 Compare the 2nd and 3rd numbers. Swap if they are not in ascending order.

3 Continue making comparisons and swapping entries until the end of the list.

4 Repeat steps 1–3 until a pass is made where no swaps are needed. Stop.

Bubble sort (cont.)

Sort these numbers into ascending order using the bubble sort:

6, 7, 1, 4, 14

First pass

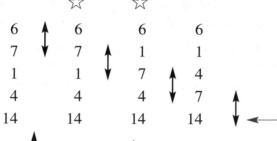

On the first pass the largest number goes to the bottom.

4 comparisons and 2 swaps ☆

Second pass

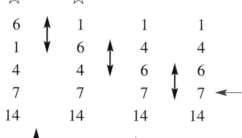

On the second pass the next largest number goes into place.

3 comparisons and 2 swaps ☆

On the third pass there will be another 2 comparisons but no swaps will be made.

Total comparisons = 10   Total swaps = 4

Maximum number of comparisons is

$$1 + 2 + ... + (n - 1) = \frac{n(n - 1)}{2}$$

Maximum number of swaps is

$$1 + 2 + ... + (n - 1) = \frac{n(n - 1)}{2}$$

Counting comparisons and swaps is important, as you may be asked to comment on the efficiency of different sorting algorithms. The fewer comparisons and swaps, the more efficient the algorithm is.

Shuttle sort

1 Compare the 1st and 2nd numbers in the list. Swap to put in ascending order, if necessary. (First pass)

2 Compare the 2nd and 3rd numbers. Swap if necessary. If swap is made, compare 2nd and 1st numbers and swap if necessary. Otherwise, go to step 3. (Second pass)

3 Compare the 3rd and 4th numbers. Swap if necessary. If swap is made, compare 2nd and 3rd numbers. Swap if necessary. If swap is made, compare 2nd and 1st numbers. Swap if necessary. At any point when no swap is made, move on to the next pass. (Third pass)

4 Continue in this way until the numbers are sorted.

You must memorise the shuttle sort rules as they will not always be given to you in the question.

For n entries there will be $n - 1$ passes.

Shuttle sort (cont.)

Sort these numbers into ascending order using the bubble sort:
6, 7, 1, 4, 14

Pass	1	2		3			4
	6	6	6	1	1	1	1
	7	7	1	6	6	4	4
	1	1	7	7	4	6	6
	4	4	4	4	7	7	7
	14	14	14	14	14	14	14
Comparisons	1	1	1	1	1	1	1
Swaps	0	1	1	1	1	0	0

Total comparisons = 7   Total swaps = 4

For this data the shuttle sort is more economical than the bubble sort, because there are 3 fewer comparisons.

The maximum number of comparisons is
$$1 + 2 + 3 + \dots + (n - 1) = \frac{n(n - 1)}{2}$$
The maximum number of swaps is
$$1 + 2 + 3 + \dots + (n - 1) = \frac{n(n - 1)}{2}$$

Quick sort

You must memorise the quick sort rules as they will not always be given to you in the question.

1 Take the 1st number in the list as a pivot.
2 Sort the remaining numbers into two sub-lists:
   (*i*) numbers smaller than the pivot
   (*ii*) numbers larger than the pivot.
   Do not re-order these sub-lists. Position the pivot between the two sub-lists.
3 If a sub-list contains more than one entry, repeat steps 1 and 2 for each sub-list until all sub-lists contain exactly one entry.

6 is the first pivot.

Step 2:
4 comparisons and 4 swaps.

Step 3:
2 comparisons and no swaps.

Sort the following numbers into ascending order using the quick sort: 6, 7, 1, 4, 14

⑥	7	1	4	14	(step 1)
1	4	⑥	7	14	(step 2)
①	4	⑥	⑦	14	(step 3)

Total comparisons = 6   Total swaps = 4

Comparing sorting methods

The bubble sort and shuttle sort have the same total number of maximum comparisons. However, the number of comparisons made for identical lists of numbers will vary from method to method.

For the given set of numbers the bubble sort was the least efficient, requiring more comparisons than both the shuttle sort and the quick sort. The quick sort was the most efficient, using the fewest comparisons.

Note that in all methods the same number of swaps were needed to get the numbers in order.

Binary search

This method is used for locating a particular entry within an ordered list (e.g. names in alphabetical order). You might find a word in the dictionary in a similar way.

1   Look at the middle item. Is this the required entry? If 'Yes', stop. If 'No' select the half list in which the item will occur.

2   Choose the middle item again. Is this the required entry? If 'Yes', stop. If 'No' select the new half list in which the entry will occur.

3   Repeat steps **1** and **2** until the required entry is found.

This is a very efficient way of finding particular entries, as each time the list you are searching is halved.

EXAMPLE

Find number 32 in this set of numbers.

28   32   40   ④④   62   68   74

The middle item is 44 – incorrect. The entry required is less than this, so look in the first half.

28   ㉜   40

The middle item in the first half is 32. This is the entry we wanted.

Bin packing

Here, in theory, we want to pack boxes of equal width and depth but varying heights into bins, which will accommodate the width and depth exactly. The bins have a maximum height, which cannot be exceeded.

These methods are required when loading a car ferry and when storing information on computer disks, for example.

There are three different algorithms for bin packing. They are called **heuristic algorithms** because they will give a good solution to any bin packing problem.

You need to be able to compare the efficiency of the solutions. You must follow each method exactly, as each method will give different combinations of boxes.

Full bin algorithm

Look for combinations of boxes that will create full bins. (This requires reasoning and comparing values, which a computer is unable to do.)

When no more full bins can be made, the remaining boxes are packed into the first available space.

Full bin algorithm (cont.)

EXAMPLE

Pack the following boxes into bins with a maximum height of 10.

Heights of boxes:

4 2 4 6 5 4 3 5 8 2

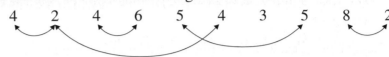

Find sets of boxes whose heights add to 10.

4   2   4   6   5   4   3   5   8   2

First fit algorithm

Take the first box in the list and find the first available space for it in the rack of bins. Continue with the 2nd, 3rd, 4th boxes, etc. until all the boxes are packed into the bins.

METHOD NOTE

Can waste a lot of space and rarely gives the optimum solution.

EXAMPLE

Pack the following boxes into bins with a maximum height of 10.

Heights of boxes: 4 2 4 6 5 4 3 5 8 2

4 goes in the 1st bin, so does 2 and the next 4. This bin is now full.

6 must go into the 2nd bin but 5 cannot fit, so it goes into the 3rd bin.

4 fits in the 2nd bin; this bin is full.

Continue in this way until all the boxes are packed.

4		2		
2	4	3		
4	6	5	5	8

First fit decreasing algorithm

1 Write the list of box sizes in decreasing order.
2 Apply the first fit algorithm.

EXAMPLE

Pack the following boxes into bins with a maximum height of 10.

Heights of boxes: 4 2 4 6 5 4 3 5 8 2

Re-order boxes: 8 6 5 5 4 4 4 3 2 2

8 goes in the 1st bin. 6 won't fit in the 1st bin so it goes into the 2nd bin.

5 won't fit in either the 1st or the 2nd bins so it goes into the 3rd bin.

Continue in this way until all the boxes are packed.

			2	
2	4	5	4	
8	6	5	4	3

Comparing solutions

In these examples the first fit decreasing and the full bin algorithms both gave the optimum solution. The first fit algorithm gave a solution that used the same number of bins but not quite so economically, i.e. empty space was split between two bins.

Algorithms

1   Write a flow chart which will calculate the radius of the largest circle with an area less than 200 square units. It should print out only the whole number radius of the largest circle.

2   Sort these numbers into ascending order using the bubble sort. Count the comparisons and swaps.

    68   74   40   62   44   28   32   46

3   Sort these numbers into ascending order using the shuttle sort. Count the comparisons and swaps.

    68   74   40   62   44   28   32   46

4   Sort these numbers into ascending order using the quick sort. Count the comparisons and swaps.

    68   74   40   62   44   28   32   46

5   Which of the sorting methods used in questions **2**, **3** and **4** was the most efficient?

Explain your answer in terms of the number of comparisons and the number of swaps.

6   A builder buys timber in 3 m lengths. He needs to cut the timber into the following lengths.

0.5 m	0.6 m	1.0 m	1.2 m	1.5 m	2.4 m
6	7	2	2	3	2

Use the first fit algorithm and the first fit decreasing algorithm to find the least number of lengths that the builder must buy.

Which gives the best result? Is either solution the optimum solution?

> **HINT**
>
> Can you find a better solution using the full bin algorithm?

7   A small vehicular ferry has 3 lanes, each 25 metres long. The vehicles listed are waiting to be loaded.

minibus	7 m	lorry	12 m	van	5.5 m
car	4 m	4 × 4	4.5 m	car	4 m
car	3 m	car	4 m	car	4 m
van	4.5 m	car	3 m	lorry	9 m
4 × 4	5 m	lorry	10 m		

Use your knowledge of bin packing to find how many vehicles can be taken on one trip.

Explain your answer. Which method was the best?

Which vehicles are left behind?

2 Graphs and networks

Vocabulary and definitions

Definitions can vary for different exam boards. Be sure you know the definitions your board uses for these words, as they could appear in exam questions.

A **graph** is a set of points (called **vertices** or **nodes**) and a collection of lines (called **edges**) joining together pairs of these points.

A **simple graph** has no loops and no more than one edge between each pair of vertices.

A **network** is a graph with numerical values or weights attached to each edge.

Degree 3 vertex.

The **degree of a vertex** is equal to the number of edges going into and out of it. This may be odd or even.

A **digraph** is a graph where each edge has a direction associated with it. (Indicated by the arrows.)

A **walk** is a sequence of edges joining vertices one after another. Repeated edges and vertices are allowed.

A **path** is a sequence of edges joining vertices one after another where no edge or vertex is repeated.

Not necessarily a direct path.

A graph is **connected** if a path exists from every vertex to every other vertex.

A graph is **complete** if every pair of vertices is connected.

See page 169 for the Hamiltonian cycle.

A **cycle** is a path which starts and finishes at the same vertex.

A **tree** is a connected graph with no cycles.

Two graphs are **isomorphic** if their sets of vertices and edges are exactly alike.

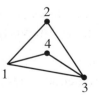
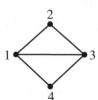

Euler tour

This is a path using all of the **edges** of the graph **exactly once**.

For a graph to contain an Euler tour it must have:
(i)   all vertices with even degree

or

(ii) *exactly* two vertices with odd degree.

REVISION NOTE

With all even vertices, an Euler tour may start and finish at any vertex.

With two odd vertices, the Euler tour must start at one odd vertex and finish at the other.

(i) All even vertices   (ii) Two odd vertices   (iii) No Euler tour

REVISION NOTE

Remembering the rules above makes it easy to see where to start and finish a tour.

Hamiltonian cycle

This is a **cycle** using all the **vertices** of the graph **exactly once**. There are no known necessary and sufficient conditions for a Hamiltonian cycle.

(i) This graph has a   (ii) This graph has no
      Hamiltonian cycle.   Hamiltonian cycle.

 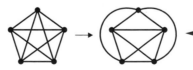

REVISION NOTE

The cycle must start and finish at the same vertex, which is found by trial and error.

Planar and non-planar graphs

A graph is **planar** if it can be drawn without any two edges crossing.

(i) A planar graph   (ii) A non-planar graph

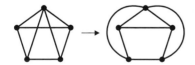 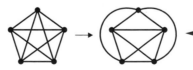

EXAM NOTE

You should be able to redraw a given planar graph to show that the edges do not cross.

Odd vertices in a graph

If a graph contains odd vertices, the number of odd vertices must be even. This is because each edge contributes 2 to the vertex order total, so the vertex order total will be even.

So there must be an even number of odd vertices, because pairs of odd numbers are needed to make the total even.

EXAMPLE

In the planar graph in (i) above, there are 1 even and 4 odd vertices. The vertex order total is $4 + 3 + 3 + 3 + 3 = 16$. Each edge contributes twice in the order total – once at each of two vertices.

Bipartite graphs and matchings

A **bipartite graph** is one where the vertices can be divided into two distinct sets. Edges can only exist if they join a vertex in one set to a vertex in the other set.

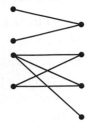

These are both planar graphs.

A **complete bipartite graph** is one where every vertex in one set is joined to every vertex in the other set.

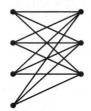

This graph is non-planar. It cannot be drawn without edges crossing.

A **matching** is a graph that links vertices from one set to another. Linked vertices are linked by exactly **one** edge. Not all vertices are linked to a vertex in the other set.

A **maximum matching** is a matching where as many pairs of vertices as possible (each pair containing one from each set), are linked by exactly one edge.

Finding a maximum matching

EXAMPLE

Find the maximum matching for this bipartite graph.

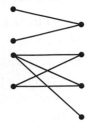

1 Draw an initial matching, selecting edges from all those available.

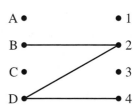

METHOD NOTE

To create an alternating path, start with an edge not in the initial matching. Follow this by an edge from the initial matching, then an edge not from the initial matching, etc.

2 Choose a non-connected vertex and look for an alternating path to another non-connected vertex.

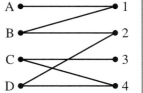

B not connected, B → 2 was not in initial matching, 2 → D was, D → 4 was not.

3 Delete any edge on the alternating path that was in the initial matching and include any edges from the initial matching that were not on the alternating path.

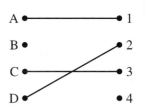

4 Repeat steps **2** and **3** until no alternating paths can be found.

EXAM NOTE

This solution looks obvious. But in an exam you are tested on following the rules, so make sure you work through the algorithm.

Minimal spanning trees

A minimal spanning tree is a connected graph where all the vertices of the graph are joined by the shortest edges. There are no cycles.

For example, a minimum spanning tree would be used to connect towns to cable television. Each town must be connected to another town with cable but not every town has to be connected to the source.

Prim's algorithm

EXAMPLE

Find a minimal spanning tree for this network using Prim's algorithm.

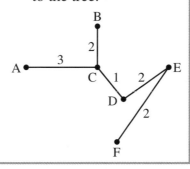

1   Choose any vertex as a starting point.   A

2   Connect it to another vertex using the shortest edge.

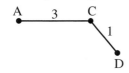

3   Choose the next shortest edge to join a new vertex to the spanning tree.

4   Repeat step **3** until all the vertices are joined to the tree.

Kruskal's algorithm

EXAMPLE

Find a minimal spanning tree for the network above using Kruskal's algorithm.

1   List all the edges in ascending order of length
    (1) CD   (2) DE, BC, EF   (3) AC, DF, CE
    (4) AB, BE   (5) AF

2   Select the smallest edge and begin to draw the spanning tree.

3   Select the next smallest edge which does not form a cycle. Add this to the spanning tree.

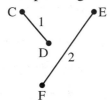

4   Repeat **3** until all the vertices are included in the spanning tree.

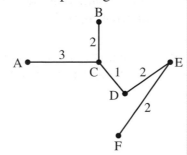

Flows in networks

A **directed graph** describes the flow of a commodity from a starting point called a **source** to a final destination called a **sink**, for example water in pipes or people on a fire escape.

A source, S, is a vertex with no input and a sink, T is a vertex with no output.

Here is an example of a directed graph with a source at S and a sink at T.

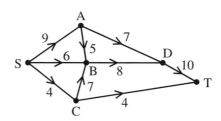

The weight attached to each edge is the maximum capacity of that edge.

When a flow is established, the input for each vertex is the same as the output for that vertex.

Maximum flow

> The maximum flow for a network is equal to the value of the minimum cut for that network.

A **cut** is a line that divides the network into two parts – one part containing the source and the other containing the sink.

Adding the capacities of all the edges crossing the cut gives the value of the cut.

EXAMPLE

Find the minimum cut for the network below and hence find any solution for the maximum flow.

Value of the left-hand cut is $9 + 6 + 4 = 19$

Value of the right-hand cut is $10 + 4 = 14$

To establish the minimum cut, calculate the values for all possible cuts.

For this network, 14 is the minimum cut, so the maximum flow will also be 14.

This network shows just one solution for maximum flow.

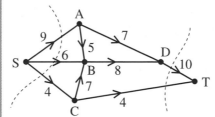

METHOD NOTE

This network shows the direction of flow and the maximum capacity for each edge.

METHOD NOTE

The flow must be towards the sink to be added into the value of the cut.

EXAM NOTE

Make sure that you test all possible cuts to get the minimum.

METHOD NOTE

There are several different maximum flow solutions for this network. This is not always the case. However, all solutions have a flow of 14 leaving the source and a flow of 14 entering the sink.

3 Shortest path problems

Dijkstra's shortest path algorithm

This algorithm is a **greedy algorithm** that chooses the shortest path from a given starting point to a given end point. Greedy algorithms take the best solution at each step, with no regard to the overall effect.

EXAMPLE

Find the shortest path from S to T for the network below.

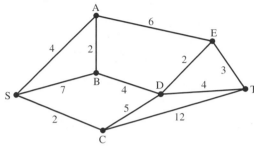

The weights on the edges are all distances. We are looking for the shortest distance from S to T.

1   At each vertex draw a box. ▭
    Write a temporary distance label at each vertex connected directly to S.
    Select the smallest (S to C). Make it permanent.
    Write 2 in the order box as it is the second vertex to be made permanent. This is the nearest vertex to S.

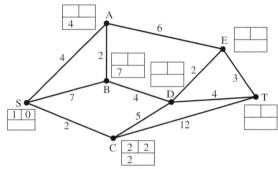

2   Write temporary labels at each vertex connected to C. If a temporary label exists, replace this label if the new distance via C is shorter.
    Select the vertex with the smallest temporary label. Make it permanent. (A with distance 4.)

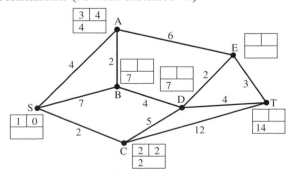

3   Continue in this way until you have built up a path from S to T. Once all vertices have a permanent label, work backwards to find which route you should take to travel the least distance.

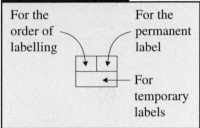

METHOD NOTE

For the order of labelling ↘   For the permanent label ↙

▭ For temporary labels

METHOD NOTE

Fill in the boxes (order of labelling and permanent distance).

EXAM NOTE

You must follows the rules precisely if you want to get full marks. The order in which you include edges is very important.

Dijkstra's shortest path algorithm (cont.)

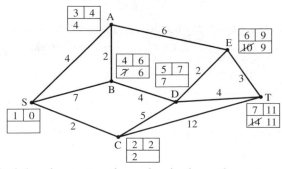

4   To find the shortest path, go back along the network from T to S, subtracting the distance along the arcs. Arcs included in the shortest path will obey the normal rules of subtraction.

e.g. $T - D$ is $11 - 4 = 7$. Therefore, D is on the shortest path because the permanent label at D is 7.

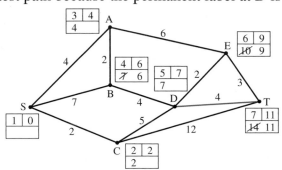

However, $T - C$ is $11 - 12 = -1$, but the permanent label at C is 2. Therefore, C is not on the shortest path.

5   The shortest path is $S - C - D - T$ and is 11 units long.

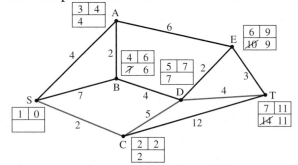

METHOD NOTE

Highlight each arc to be included in the shortest path.

EXAM NOTE

Always state your final solution in the correct order from start to finish as well as showing it on the network.

METHOD NOTE

For a vertex with even degree, each time you travel into that vertex you have a way of getting out of it again. So you should never get stuck with no way out as you travel around the network, until you come back to the start.

The Chinese postman problem (route inspection)

The aim is to travel along every arc exactly once and end up back where you started. Think of this as putting your pencil on one vertex and drawing over every edge exactly once without taking your pencil off the paper. This can be done if all vertices are even.

EXAMPLE

Find a route around this network that uses every edge exactly once.

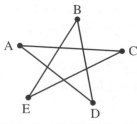

A quick check shows that all vertices are even, so we can start anywhere.

Starting at A move to C then E then B then D then A. This route covers all edges exactly once and is therefore a route inspection solution for the given network.

The Chinese postman problem (route inspection) (cont.)

When there are two or more odd vertices, extra edges must be added to make all the vertices even. This will make it possible to find a route.

EXAMPLE

Find a route around this network.
Note the degree of all vertices.

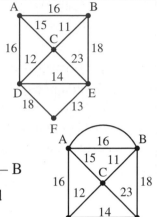

A : 3	D : 4
B : 3	E : 4
C : 4	F : 2

There are two odd vertices. Start at one odd vertex and finish at the other.

A – B – C – D – E – C – A – D – F – E – B

Then, add an edge between the two odd vertices (AB) to get back to the start.

Total length of route is 172 units.

METHOD NOTE

Find the route by trial and error.

METHOD NOTE

Adding this arc makes all the vertices even and hence a solution can be found.

The length of the new edge between A and B is the same as the length AB.

EXAMPLE

Find an inspection route around this network.

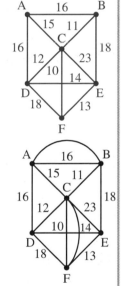

Here there are more than two odd vertices. Take the odd vertices in pairs and calculate the distances between them.

AB & CF = 16 + 10 = 26

AC & BF = 15 + 21 = 36

AF & BC = 25 + 11 = 36

Select the pair with the lowest total distance, in this case AB & CF.

Each of these routes A – B and C – F will be repeated in the final route.
Add edges for A – B and C – F.

The solution will be

A – B – C – D – E – F – C – F – D – A – C – E – B – A

Total distance = 192 units

METHOD NOTE

Adding these arcs makes all the vertices even, so a route can be found.

Rules for the Chinese postman problem

- If all vertices are even, start anywhere.
- If there are two odd vertices, start at one of them, finish at the other, and add in the shortest extra edge between these vertices.
- If there are more than two odd vertices, find the shortest edges to repeat. Add these to your network. Now all vertices are even, so start anywhere.

Networks and shortest path problems

1   Workers   A   are able to do jobs   2, 3
            B   1, 3, 5
            C   2, 4
            D   3, 5
            E   1, 5
                                            respectively.

   Draw this relationship as a bipartite graph and find a maximum matching.

2   Using Prim's or Kruskal's algorithm, find a minimal spanning tree for the network below.

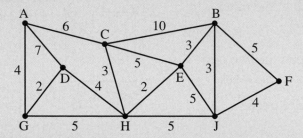

3   Using Dijkstra's algorithm, find the shortest path from A to F for the network in question **2**.

4   Using the rules for the Chinese postman problem, find the best route around this network.

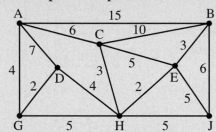

> **HINT**
> Look for the vertices with odd degree first.

5   Find the maximum flow from S to T for this network.

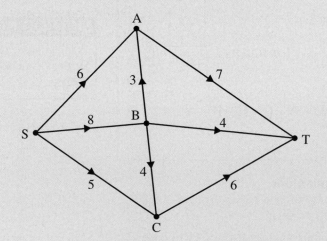

> **HINT**
> Look at all the cuts to find the minimum.

4 Critical path analysis

This type of network analysis is used to find the minimum time needed to complete a project consisting of many activities. Some of these activities may be done simultaneously. The longest path through the network will include all the critical activities.

> An activity is **critical** if a delay in finishing it will delay the whole project.

Precedence networks

A **precedence network** is a diagram showing how the activities in the table are linked to each other.

EXAMPLE

From the table shown, draw a precedence network and find the critical path for the project.

Activity	Duration (days)	Preceding activity	No. of workers
A	1.5		1
B	5		1
C	1	A, B	2
D	2	A	1
E	3	C	1
F	0.5	C	1
G	1.5	E, D	1

1 Activities for the project must be listed along with their immediately preceding activities, their duration and the number of workers required.

2 Arcs, labelled with an activity and duration, make up the precedence network. Vertices represent events – the start or finish of an activity.

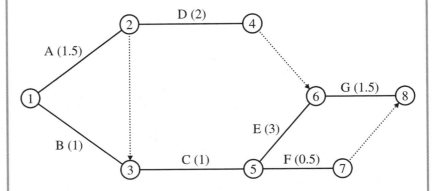

Dotted arcs are dummy arcs showing precedence relations with no time duration. For example: C needs both A and B but D needs only A. A dummy arc is drawn from 2 to 3 to show that C cannot start until A is finished.

EXAM NOTE

Make sure your labelling is clear.

Earliest start time
for next activity.
(Follow all paths into
this vertex and write
the longest in the box.)

↓

Latest start time
for next activity.
(The longest path
back to this activity
from the next activity.)

REVISION NOTE

A forward pass finds the **earliest** start time for each activity.

3 Perform a forward pass to calculate the earliest start time for each event.

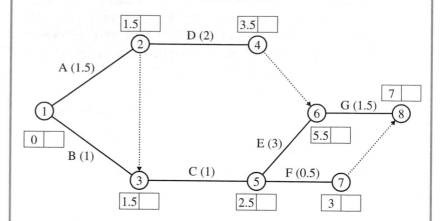

If there is more than one path to a vertex, take the longest time. For example, the earliest start time for C is 1.5, because C cannot start until A has finished.

Earliest start time for E equals earliest start time for C plus duration of C.

4 Perform a backward pass to calculate the latest start time for each event.

Latest start time for G = (7.0 − duration of G) = 5.5.

5 The **critical path** contains those events where earliest and latest start times are equal. This is highlighted here. Any delay in these activities will delay the whole project.

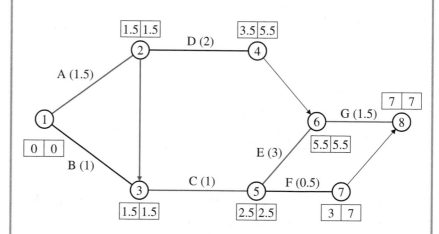

Critical activities are A, C, E and G.

Shortest completion time is 7 days.

REVISION NOTE

Slack time is spare time associated with non-critical activities.

Resource allocation shows the fluctuations in the number of workers required for the project.

Cascade charts

From the information on the precedence network we can draw a cascade chart. This relates the activities to a time scale, allows **slack time** to be seen and enables **resource allocation**.

Draw a cascade chart for the network on page 178.

First we set out all the tasks, beginning at their earliest start time.

Where some slack time exists for an activity, dotted lines extend beyond the end of that activity to the time by which they must be finished.

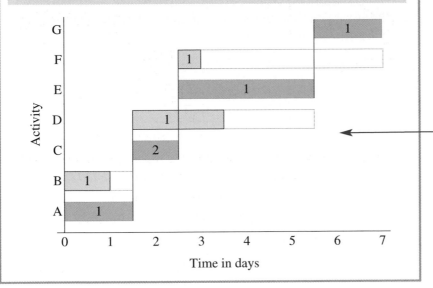

The dotted lines indicate the length of slack time available for the activity, should it be needed.

The darker shaded activities are critical activities and have no slack time.

Numbers in boxes show numbers of workers needed for the activity.

Activities B, D and F may be delayed if necessary. This may be useful if there are not enough workers to complete these activities at the early start times.

Resources histograms

To draw the resources histogram, draw a block with length the length of the activity and height the number of workers required.

Put critical activities in one colour along the horizontal axis. Then put non-critical activities above them, in the correct places for their given starting times.

The first resources histogram shows that we need 3 workers some of the time.

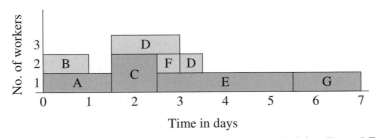

By making use of the slack time available in activities D and F we can finish this project in 7 days using at most 2 workers.

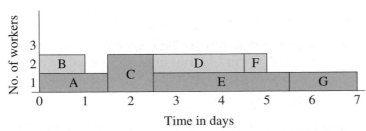

If D starts after C has finished and F starts after D has finished, we only need two workers. Activity D could be delayed even longer, provided it is finished by Day 5.5. Activity F can also be delayed longer, provided it is finished by Day 7.

Activity D is delayed until day 2.5 and will be finished before activity G starts. Activity F is delayed by 2 days but will be finished before the end of the project. Only 1 worker is needed for the last two days.

5 Linear programming

This is a technique used for finding the best solution to a problem, given a variety of constraints.

Graphical solution using vertices

EXAMPLE

A baker makes two types of choc-chip cookies – standard and extra-choc. The ingredients for making a batch of each type are shown in the table.

	Standard (x)	Extra-choc (y)
Flour	8 kg	8 kg
Sugar	2 kg	1 kg
Choc chips	1 kg	3 kg

The baker has 64 kg of flour, 14 kg of sugar and 18 kg of choc chips.

Standard cookies make a profit of £25 per batch and extra-choc cookies make a profit of £15 per batch.

How many batches of each should the baker make to maximise the profit?

1   Formulate the **constraints** and **objective function**.

$8x + 8y \leq 64$ (flour)
$2x + y \leq 14$ (sugar)
$x + 3y \leq 18$ (choc chips)

The objective function is the profit. We want $25x + 15y$ to be a maximum, for $x \geq 0$, $y \geq 0$

2   Draw the graphs of the constraints. Shade out the area *not* required.

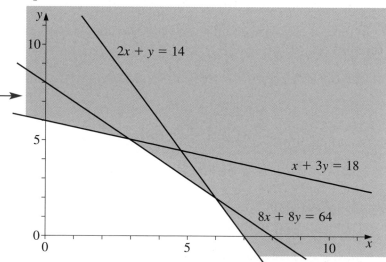

3   Find the points of intersection of the constraints, as the maximum profit will occur at one of these.

By solving simultaneous equations, we find these points are:
(0, 6) (3, 5) (6, 2) (7, 0)

4   Checking each of these values in the objective function, we find that the maximum profit is £180. This occurs when the baker makes 6 batches of standard cookies and 2 batches of extra-choc cookies.

METHOD NOTE

The variables here are standard and extra-choc. They should be labelled x and y. This makes the inequalities easier to formulate and graph.

EXAM NOTE

Read linear programming problems carefully, as you may have to work out the inequalities for the constraints yourself.

The **constraints** in this example are the amounts of ingredients available.

The **objective function** is the constraint for which we must find a maximum.

METHOD NOTE

The unshaded area of the graph is called the **feasible area**. All solutions will be in here.

METHOD NOTE

The optimum solution will be at one of the **vertices** of the feasible area. You need to check the coordinates of each corner in the objective function.

For example, solving
$8x + 8y = 64$ and $2x + y = 14$
gives $x = 6$, $y = 2$

For example:
Substitute $x = 3$ and $y = 5$ into
$25x + 15y$
$25 \times 3 + 15 \times 5 = 150$

Linear programming and integer solutions

When integer values are required for the solution to a linear programming problem, the vertices of the feasible area may not yield a suitable result. This may occur when you are given a problem on manufacturing.

REVISION NOTE

The graphs of the constraint inequalities may not cross at integer values.

EXAMPLE

Given the constraints:

$$8x + 8y \leqslant 600$$
$$2x + y \leqslant 140$$
$$x + 3y \leqslant 180$$
$$x \geqslant 0, \ y \geqslant 0$$

find a maximum value for $P = 15x + 20y$ for integer x and y.

EXAM NOTE

Always draw graphs and shade the area not required.

Draw the graphs to find the feasible area.

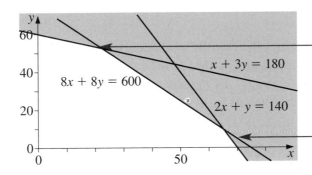

METHOD NOTE

Solve the simultaneous equations
$$8x + 8y = 600$$
$$x + 3y = 180$$
to find the coordinates of this vertex.

Solve the simultaneous equations
$$8x + 8y = 600$$
$$2x + y = 140$$
to find the coordinates of this vertex.

Find the vertices of the feasible area by solving simultaneously the equations corresponding to the constraints.

The vertices are:

$$(0, 60) \quad (22.5, 52.5) \quad (65, 10) \quad (70, 0)$$

Checking these in the equation of P gives:

P is a maximum of 1387.5 when $x = 22.5$ and $y = 52.5$.

However, **x and y must be integer values**, so look at all possible integer solutions **close to** the vertex that gives the maximum solution.

Possible solutions are:

$$(21, 53) \quad (22, 52) \quad (23, 52) \quad (23, 51)$$

$(21, 53)$ is on the $x + 3y = 180$ line with value 1375

$(22, 52)$ is inside the feasible area with value 1370

$(23, 52)$ is on the $8x + 8y = 600$ line with value 1385

$(23, 51)$ will be less than 1385, so discard this solution.

The integer solution is $x = 23$ and $y = 52$, giving a maximum value of $P = 1385$.

METHOD NOTE

You must check that these are all in the feasible area before looking for the optimum solution. You can do this if your graph is well drawn and accurate. Otherwise, check that each pair satisfies all the constraints.

The Simplex method

This method also finds the optimum solution to a linear programming problem but does not involve drawing graphs. It is especially useful for constraints that have 3 or more variables and also when we are interested in the amount of materials that may be left unused when the best solution is found.

This is the same problem as on page 180.

EXAMPLE

Find a maximum value of P given these constraints:

$$8x + 8y \leqslant 64 \qquad 2x + y \leqslant 14 \qquad x + 3y \leqslant 18$$

$$P = 25x + 15y$$

1. Write the inequalities as equations by introducing **slack variables**. Rewrite the objective function equal to 0.

$$8x + 8y + r = 64$$
$$2x + y \ + s = 14$$
$$x + 3y + t = 18$$
$$P - 25x - 15y = 0$$

METHOD NOTE

r, s and t are all **slack variables**.

2. Construct the **Simplex tableau**. The first column is the pivot column (shaded in the tableau below). Divide each value entry by the corresponding pivot column entry. **The smallest result indicates the pivot row.**

METHOD NOTE

The tableau shows all the constraints and the objective function in a table for ease of calculations.

x	y	r	s	t	P	Value	
8	8	1	0	0	0	64	**(64 ÷ 8 = 8)**
2	1	0	1	0	0	14	**(14 ÷ 2 = 7)**
1	3	0	0	1	0	18	**(18 ÷ 1 = 18)**
−25	−15	0	0	0	1	0	

Pivot row →

Smallest value indicates pivot row, after dividing by 2.

3. Divide each entry in the pivot row by the pivot column value.

x	y	r	s	t	P	Value	
8	8	1	0	0	0	64	**R1**
1	0.5	0	0.5	0	0	7	**R2**
1	3	0	0	1	0	18	**R3**
−25	−15	0	0	0	1	0	**R4**

Shows new values of pivot row.

4. Use the pivot row to create zero entries in all rows in the pivot column, except in the pivot row.

x	y	r	s	t	P	Value	
0	4	1	−4	0	0	8	**R1 − 8 × R2**
1	0.5	0	0.5	0	0	7	**R2**
0	2.5	0	−0.5	1	0	11	**R3 − R2**
0	−2.5	0	12.5	0	1	175	**R4 + 25 × R2**

Pivot row →

R3 − R2 will give a zero in the first column of row 3.

The Simplex method (cont.)

5 The second column now becomes the pivot column. Divide each value entry by the corresponding pivot column entry.

x	y	r	s	t	P	Value	
0	4	1	−4	0	0	8	(8 ÷ 4 = 2)
1	0.5	0	0.5	0	0	7	(7 ÷ 0.5 = 14)
0	2.5	0	−0.5	1	0	11	(11 ÷ 2.5 = 4.4)
0	−2.5	0	12.5	0	1	175	

METHOD NOTE

Divide each value entry by the y entry in the same row.

Pivot row

Smallest value indicates the pivot row.

6 Divide each entry in the new pivot row by the pivot column entry.

x	y	r	s	t	P	Value	
0	1	0.25	−1	0	0	2	**R1 ÷ 4**
1	0.5	0	0.5	0	0	7	
0	2.5	0	−0.5	1	0	11	
0	−2.5	0	12.5	0	1	175	

7 Use the new pivot row to create zero entries in all rows in the pivot column, except in the pivot row.

x	y	r	s	t	P	Value	
0	1	0.25	−1	0	0	2	**R1**
1	0	−0.125	1	0	0	6	**R2 − 0.5 × R1**
0	0	−0.625	2	1	0	6	**R3 − 2.5 × R1**
0	0	0.625	10	0	1	180	**R4 + 2.5 × R1**

y is the only variable left in the first row. r, s and t are slack variables.

This will give a zero in the second column of the second row.

If there are three variables in your constraints, you may have to let the third column be a pivot column as well and go through the steps again.

8 The final solution is found when all entries in the last row are positive. (Notice that there is only one entry of 1 in each of the x and y columns.)

The final solution is read from the tableau as follows:

Maximum value for the objective function is £180. This occurs when $x = 6$ and $y = 2$. (These values are in the same rows as the 1 entries for x and y.)

We found the same solution in the earlier example on page 178. By substituting $x = 6$ and $y = 2$ into the tableau equations we see that $r = 0$, $s = 0$ and $t = 6$.

This means that when making the cookies on page 178, both the flour and sugar were completely used up but 6 kg of choc chips remained unused.

$$8x + 8y + r = 64$$
$$8 \times 6 + 8 \times 2 + r = 64$$
$$r = 0$$

6 Boolean algebra

Propositional logic symbols

REVISION NOTE

You need to understand the meaning of all the symbols and to be able to combine propositions using **truth tables**.

All arguments consist of some basic statements called **propositions**. A proposition is a meaningful sentence that may be either **true** or **false**. In Boolean algebra we use symbols to represent these propositions and the relationship between them.

For example,
p: it is raining.
~p: it is not raining.

> If **p** is a proposition then ~**p** represents 'not p'.
>
> If **p** and **q** are both propositions then **p** ∧ **q** represents 'p and q'.
>
> If **p** and **q** are both propositions then **p** ∨ **q** represents 'p or q or both p and q'.

Symbols and propositions may be combined to make compound propositions.

Truth tables

Truth tables are used to make sense of compound propositions. A truth table looks at all possible combinations of true and false. The most common truth tables are:

1 and 0 are used in truth tables to represent true and false respectively.

AND			**OR**			**NOT**	
a	b	a ∧ b	a	b	a ∨ b	a	~a
0	0	0	0	0	0	0	1
0	1	0	0	1	1	1	0
1	0	0	1	0	1		
1	1	1	1	1	1		

REVISION NOTE

If you cannot remember these tables, you need to be able to work them out.

a ⇔ b is the same as a ⇒ b and a ⇐ b together.

IMPLIES			**IS IMPLIED BY**			**EQUIVALENT TO**		
a	b	a ⇒ b	a	b	a ⇐ b	a	b	a ⇔ b
0	0	1	0	0	1	0	0	1
0	1	1	0	1	0	0	1	0
1	0	0	1	0	1	1	0	0
1	1	1	1	1	1	1	1	1

A **tautology** is a compound statement that is always true.

REVISION NOTE

Remember the definition of a tautology and use a truth table to prove it if necessary.

EXAMPLE

Show that (**a** ∧ **b**) ⇒ **a** is a tautology (it is always true).

a	b	a ∧ b	a	(a ∧ b) ⇒ a
0	0	0	0	1
0	1	0	0	1
1	0	0	1	1
1	1	1	1	1

All combinations are true, therefore this is a tautology.

Truth tables (cont.)

A **contradiction** is a compound statement that is always false.

EXAMPLE

Show that $a \wedge (b \wedge \sim a)$ is a contradiction (it is always false).

a	b	~a	b ∧ ~a	a	a ∧ (b ∧ ~a)
0	0	1	0	0	0
0	1	1	1	0	0
1	0	0	0	1	0
1	1	0	0	1	0

All outcomes are false, so this is a contradiction.

METHOD NOTE

When using a truth table, always start with every possible combination of propositions. Work across the rows.

Laws of Boolean algebra

Associative	$a \vee (b \vee c) = (a \vee b) \vee c$
	$a \wedge (b \wedge c) = (a \wedge b) \wedge c$
Distributive	$a \wedge (b \vee c) = (a \wedge b) \vee (b \wedge c)$
	$a \vee (b \wedge c) = (a \vee b) \wedge (b \vee c)$
Commutative	$a \vee b = b \vee a$
	$a \wedge b = b \wedge a$

REVISION NOTE

You could use truth tables to prove each of these. This has been done for De Morgan's laws below.

De Morgan's laws

$\sim(a \vee b) = \sim a \wedge \sim b$

$\sim(a \wedge b) = \sim a \vee \sim b$

Truth table for De Morgan's law $\sim(a \vee b) = \sim a \wedge \sim b$

a	b	a ∨ b	~(a ∨ b)	~a	~b	~a ∧ ~b
0	0	0	1	1	1	1
0	1	1	0	1	0	0
1	0	1	0	0	1	0
1	1	1	0	0	0	0

METHOD NOTE

Left-hand side equals right-hand side, therefore the law holds.

Truth table for De Morgan's law $\sim(a \wedge b) = \sim a \vee \sim b$

a	b	a ∧ b	~(a ∧ b)	~a	~b	~a ∨ ~b
0	0	0	1	1	1	1
0	1	0	1	1	0	1
1	0	0	1	0	1	1
1	1	1	0	0	0	0

METHOD NOTE

Left-hand side equals right-hand side, therefore the law holds.

Switching circuits

Boolean expressions and truth tables may be represented by networks of switches that control the flow of electricity or perhaps water through a circuit or a series of pipes.

A **series circuit** with switches A and B looks like this:

The truth table for the series circuit is:

A	B	Output
0	0	0
0	1	0
1	0	0
1	1	1

There is no current flowing if A or B or both A and B are open.

Current flows when both A and B are closed.

Input	Output	
A B	A ∧ B	For a series circuit

A **parallel circuit** with switches A and B looks like this:

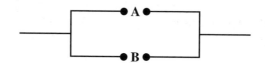

The truth table for the parallel circuit is:

A	B	Output
0	0	0
0	1	1
1	0	1
1	1	1

There is no current flowing if both A and B are open.

Current flows when A or B or both are closed.

Input	Output	
A B	A ∨ B	For a parallel circuit

EXAMPLE

Find the Boolean algebra expression for this switching circuit and give the truth table.

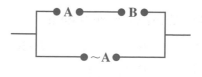

A	B	~A	A ∧ B	~A ∨ (A ∧ B)
0	0	1	0	1
0	1	1	0	1
1	0	0	0	0
1	1	0	1	1

Combinatorial circuits

Here **logic gates** act as switches for the circuits. Special symbols are used for each type of gate.

The **NOT gate** reverses the input.

Input	Output
a	**~a**
0	1
1	0

The **AND gate** gives one output from two inputs

Input		Output
a	**b**	**a ∧ b**
0	0	0
0	1	0
1	0	0
1	1	1

Output can only be 1 when **a** and **b** are both 1.

The **OR gate** gives one output from two inputs

Input		Output
a	**b**	**a ∨ b**
0	0	0
0	1	1
1	0	1
1	1	1

Output can be 1 when either **a** or **b** or both are 1.

Truth tables are used to decide the outcomes of combinatorial circuits.

EXAMPLE

Use a truth table to decide the outcomes for this logic gate diagram.

First label the circuits using your knowledge of the gates.

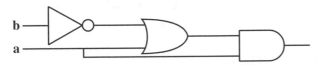

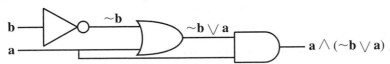

Then draw up a truth table.

a	**b**	**~b**	**(~b ∨ a)**	**a ∧ (~b ∨ a)**
0	0	1	1	0
0	1	0	0	0
1	0	1	1	1
1	1	0	1	1

This shows that the circuit is on provided that **a** is on. So **a** ∧ (~**b** ∨ **a**) simplifies to **a**.

Miscellaneous questions

1 Draw a precedence network for this project:

Activity	Preceding activity	Duration (hours)
A		8
B	A	4
C		4
D	C, E	2
E	F	5
F	B	12
G	F	2

2 Find the critical path for this network. (Times are given in days.)

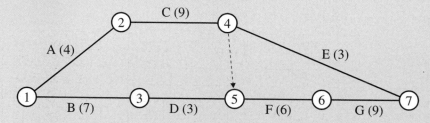

3 Draw a cascade diagram for the network in question **2**.

4 Draw a resources histogram for the information in the table below, using your solutions to questions **2** and **3**.

By how many days will this project be delayed if there are only two workers available at any time?

Activity	Number of workers required
A	2
B	1
C	1
D	2
E	1
F	2
G	1

5 Solve the following for an optimum solution

$2x + y \leqslant 8$

$x + y \leqslant 5$

$x > 0, \ y > 0$

Maximise $P = 3x + 2y$.

(*a*) Use a graphical method.

(*b*) Add slack variables s and t and use the Simplex method.

6 Is $[\mathbf{a} \wedge (\mathbf{b} \vee \sim\mathbf{b})] \Leftrightarrow \mathbf{a}$ a tautology or a contradiction?

Answers and hints to solutions – Decision and Discrete Maths

Exercise 1: Algorithms

1

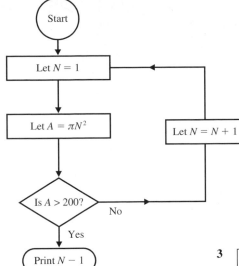

2

1st	2nd	3rd	4th	5th	6th
68	40	40	40	28	28
40	62	44	28	32	32
62	44	28	32	40	40
44	28	32	44	44	44
28	32	46	46	46	46
32	46	62	62	62	62
46	68	68	68	68	68
74	74	74	74	74	74
Total comparisons = 27; Swaps = 20					

3

1st	2nd	3rd	4th	5th	6th	7th
68	40	40	40	28	28	28
74	68	62	44	40	32	32
40	74	68	62	44	40	40
62	62	74	68	62	44	44
44	44	44	74	68	62	46
28	28	28	28	74	68	62
32	32	32	32	32	74	68
46	46	46	46	46	46	74
Total comparisons = 25; Swaps = 20						

4

1st	2nd	3rd	4th
40	28	㉘	㉘
62	32	32	32
44	㊵	㊵	㊵
46	62	44	㊸
28	44	46	46
32	46	㊷	㊷
㊸	㊸	㊸	㊸
74	74	74	74
Total comparisons = 16; Swaps = 10			

5 Quick sort was the most efficient with 16 comparisons and 10 swaps.
The shuttle sort was second best and the bubble sort was least efficient.

6

0.5							
0.5							
0.5	0.6						
0.5	0.6						
0.5	0.6	1					
0.5	0.6	0.6	1.2	1.5	1.5		
0.5	0.6	0.6	1	1.2	1.5	2.4	2.4

First fit algorithm 8 lengths,
3.1 m wasted

						0.5	
					0.6	0.5	
				0.6	0.6	0.5	
0.6	0.6	1.5	1.2	1	0.6	0.5	0.5
2.4	2.4	1.5	1.5	1.2	1	0.6	0.5

First fit decreasing 8 lengths,
3.1 m wasted

						0.5	
					0.6	0.6	0.5
				0.5	0.6	0.6	0.5
0.6	0.5	1.5	1	0.6	0.6	0.5	
2.4	2.4	1.5	1.5	1.2	1.2	1	

Full bin algorithm 7 lengths,
0.1 m wasted

Both give the same answer which is not the optimum solution.
The full bin algorithm gives the optimum solution.

7 The full bin algorithm gives the best solution. All can go except one 4 × 4.

Exercise 2: Networks and shortest path problems

1

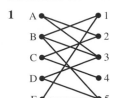 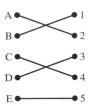 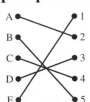 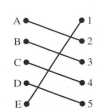

Bipartite graph and three possible solutions.

2 Minimal spanning tree

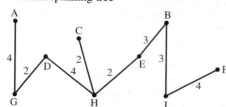

3

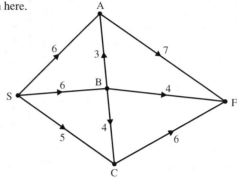

4 D, G, H, J all have odd degree.

Possible pairs are: DG & HJ: $2 + 5 = 7$;   DH & GJ: $4 + 10 = 14$;   DJ & GH: $9 + 5 = 14$.

DG & HJ give the shortest extra distance; total distance is $81 + 7 = 88$

5 Minimum cut $= 17$ (AF, BF, CF). Maximum flow $= 17$. One solution is shown here.

Exercise 3: Miscellaneous questions

1

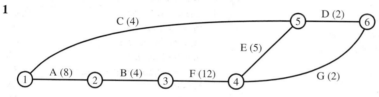

2 Solution: $A - C - F - G$; total length $= 28$ days

3

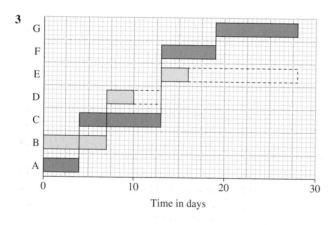

4

Whole project will be delayed for 3 days because D will have to be done after C but before F, thereby adding 3 days to the critical path. E must be moved to start at time 19 days.

5 (a) and (b) $x = 3$, $y = 2$, maximum $P = 13$.

x	y	s	t	P	Value
1	0	1	−1	0	3
0	1	−1	2	0	2
0	0	2.5	1	1	13

6 Tautology, because all the outcomes are true.

a	b	~b	b ∨ ~b	a ∧ (b ∨ ~b)	[a ∧ (b ∨ ~b)] ⇔ a
0	0	1	1	0	1
0	1	0	1	0	1
1	0	1	1	1	1
1	1	0	1	1	1

Index

$\lim\limits_{\delta x \to 0}$ 38

Σ notation   28

A

$a = bx$ 62
$a \cos x + b \sin x$ 36
acceleration   123
acceleration due to gravity   131
addition of vectors   121
algorithms   162
ambiguous case for sine rule   33
AP   28
arbitrary constant   46, 47
arc length (for a circle)   35
area between a graph and x-axis   48
   enclosed between curves   50
   of a sector   35
   under a graph   48
arithmetic progression   28
asymptotes   17, 18, 20

B

backward pass   178
behaviour for large x 20
bin packing   165
binary search   165
binomial distribution   97
bipartite graph   170
bivariate data   114
Boolean algebra   184
box plot   80
bubble sort   162

C

cascade chart   178
chain rule differentiation   53, 65, 66
change of sign methods   68
Chinese postman problem   174
class interval   83
cluster sampling   112
cobweb diagram   71
coding   85
codomain   15
coefficient of friction   144
collisions   150
combinations   87
combinatorial circuits   187
common difference   28
common ratio   28
complement   91
complete graph   168
completing the square   25
components   122, 130, 141
composite function   15, 53, 65
compound angle formulae   36
conditional probability   92
connected graph   168
connected particles   133
constant acceleration   123
constant acceleration formulae   123, 126
constant of integration   46
constraints   180
contact force   134
continuous data   82

continuous random variable   108
contradiction   185
convergence of iteration   70
$\cos x$ 31
$\operatorname{cosec} x$ 31
cosine rule   33, 34
$\cot x$ 31
critical path analysis   177
cubic graph   17
cumulative distribution function   95
cumulative frequency   84
cycle   168

D

De Morgan's laws   185
deceleration   125
decimal search   68
definite integration   48
degree   168
derivative   38
differentiation   38
digraph   168
Dijkstra's algorithm   173
direction   121, 123
discrete data   79
   random variables   94
   uniform distribution   103
discriminant   26
displacement   126
distance between two points   23
divergence of iteration   70
domain   15
double angle formulae   36
dummy arcs   177
dynamics   130, 138

E

e^x 63
edges   168
equation of motion   131
equation of trajectory   157
equilibrium   138
error bound   69
Euler tour   169
even function   16, 31
expectation   95, 97, 101, 103, 105
   algebra   105
   functions   105
   frequencies   98
exponential functions   63, 65
   series   63
external forces   138

F

$F = ma$ 131
factor theorem   27
first fit algorithm   166
first fit decreasing algorithm   166
fixed-point methods   70
flow in network   172
flow charts   162
force   130
   diagrams   130
forward pass   177
frequency density   83
friction   130, 144
full bin algorithm   166

function   15
   of a function   53

G

g (acceleration due to gravity)   131
geometric distribution   103
geometric progression   28
GP   28
gradient of a line   17, 22, 38, 40, 57
graph (discrete maths)   168
graphs and networks   168

H

Hamiltonian cycle   169
histogram   83
hypothesis testing   99

I

impulse   149
inclined plane   140
indefinite integral   47
independence   91, 92
indices   26, 62
inflexion   44
integration   46, 65
intercept   17, 20, 22
internal force   134
interquartile range   80, 84
intersection   90
interval bisection   69
inverse function   16
inverse trigonometric function   32
isomorphic   168
iteration   69

K

kinematics   123
Kruskal's algorithm   171

L

Lami's theorem   140
laws of logarithms   62
limit notation   38
limits of integration   48
linear programming   180
linear regression   115
linked variables   54
$\ln x$ 63
local maximum   42
local minimum   42
$\log_b a$ 62
logarithms   62, 64, 65
logic gates   187
lower quartile   80, 84

M

magnitude   121, 122
mapping   15
mass   131
matching   170
maximum
   flow   172
   height of projectile   156
   matching   170
   on a curve   42
   range of projectile   156

mean 79, 82
median 79, 82, 84
midpoint 23
minimal spanning tree 171
minimum on a curve 42
mode 79, 82
model (mathematical) 124, 132
momentum 149
mutually exclusive 91

N

natural logarithms 63
negative exponential function 64
network 168
Newton 130, 131
Newton's
 laws of motion 131
 second law 131, 132
 third law 133
Newton–Raphson method 72
nodes 168
non-planar graph 169
non-random sample 112
normal
 distribution 108
 reaction 140
 to a curve 41
numerical integration 73
numerical methods 68

O

objective function 180
odd function 16, 31
one-tailed test 99
ordering 87
ordinate 49

P

parallel circuits 186
path 168
Pearson's product moment
 correlation 114
periodic function 16, 31
permutations 87
perpendicular lines 22
planar graph 169
point of inflexion 44
Poisson distribution 101
polynomial 27
position 123

powers and indices 26
Prim's algorithm 171
product rule (differentiation) 56, 66
projectiles 152
propositions 184

Q

quadratic
 equation 25
 formula 25
 graph 17
quartile 80, 84
quick sort 164
quota sampling 112
quotient rule (differentiation) 58, 66

R

$r \cos (x + c)$ 36
radian measure 35, 38
random sample 113
range (of a function) 15
range (statistics) 79, 82
range of projectile 155
resolving vectors 122, 140
resources histogram 179
roots of an equation 25

S

sampling 112
scalar 121
$\sec x$ 31
second derivative 38
semi-interquartile range 84
series 28
series circuits 186
shuttle sort 163
sigma notation 28
Simplex method 182
Simplex tableau 182
Simpson's rule 73
$\sin^{-1} x$ 32
$\sin x$ 31
sine rule 33, 34
sink 171
slack variables 182
solid of revolution 51
sorting algorithms 162
source 172
speed 124
staircase diagram 71

standard deviation 79, 82
standard normal distribution 109
statics 130, 138, 140
stationary points 42
stem and leaf plots 80
straight line 17, 22
stratified sampling 113
stretch of a graph 19
subtraction of vectors 121
sum of independent Poisson
 distributions 101
sum to infinity of a GP 29
systematic sampling 112

T

$\tan^{-1} x$ 32
$\tan x$ 31
tangent to a curve 38, 40
time of flight 155
transformations of graphs 19
translation of graphs 19
trapezium rule 73
tree 168
trigonometric functions 31
trigonometric formulae 36
truth table 184, 186
two-tailed test 100

U

union 90
unit vectors 122
upper quartile 80, 84

V

variance 79, 82, 95, 97, 101, 103, 105
vector 121
 addition 121
 notation 121
 subtraction 121
velocity 123
Venn diagram 90
vertices (graph) 168
volume of a solid of revolution 51

W

walk 168
weight 131

Y

$y = mx + c$ 17, 22